Risk Analysis and the Security Survey

THIRD EDITION

Risk Analysis and
the Security Survey
THIRD EDITION

James F. Broder, CFE, CPP, FACFE

Vincennes University
Shake Learning Resources Center
Vincennes, In 47591-9986

ELSEVIER

AMSTERDAM • BOSTON • HEIDELBERG • LONDON
NEW YORK • OXFORD • PARIS • SAN DIEGO
SAN FRANCISCO • SINGAPORE • SYDNEY • TOKYO
Butterworth-Heinemann is an imprint of Elsevier

Butterworth-Heinemann is an imprint of Elsevier
30 Corporate Drive, Suite 400, Burlington, MA 01803, USA
Linacre House, Jordan Hill, Oxford OX2 8DP, UK

⊖ Recognizing the importance of preserving what has been written, Elsevier prints
its books on acid-free paper whenever possible.

Library of Congress Cataloging-in-Publication Data
Application submitted

363.11
B864r
2006

British Library Cataloguing-in-Publication Data
A catalogue record for this book is available from the British Library.

ISBN 13: 978-0-7506-7922-0
ISBN 10: 0-7506-7922-0

For information on all Butterworth–Heinemann
visit our Web site at www.books.elsevier.com

Printed in the United States of America
06 07 08 09 10   10 9 8 7 6 5 4 3 2 1

To my wife, Carolyn Oliver-Broder,
and her true love, Mr. Murphy

"If you don't know where you're going,
any road will get you there."

Lewis Carroll
Alice in Wonderland

Table of Contents

Acknowledgments

The second edition of *Risk Analysis and the Security Survey* was published in 1999. The book continues to be widely accepted within both the security profession and the academic community worldwide. However, like many security textbooks, parts of this book became outdated as a result of the events in New York City on September 11, 2001.

In consultation with Mark A. Listewnik, Acquisitions Editor–Forensics and Security, Elsevier Butterworth-Heinemann, we proposed to the publisher the writing of a 3rd edition. We suggested dividing the text into two parts: Part I, The Treatment and Analysis of Risk, and Part II, Emergency Management and Business Continuity Planning. Among other things, we wanted to highlight some of the experiences gained as a result of 9/11, as well as some of the changes in risk methodologies that resulted from the Homeland Security Act.

The largest and perhaps most important part of this project was the task of updating and rewriting Part II. This burden was graciously accepted by my friend and coauthor, Gene Tucker, CPP, of Orinda, California. I am aware of no other person in the security industry who can match his intellect, credentials, and qualifications. This is especially true in the areas of emergency management and business continuity planning. This part of the text addresses the importance of lessons learned in these areas as a result of 9/11 and more.

In the days, months, and years after September 11, 2001, we had numerous conversations with our security colleagues, emergency response personnel, and business continuity professionals. These discussions centered on the issues of "lessons learned" and the "impossibility of predicting such a tragedy." It is our opinion, which is addressed in this text, that this tragedy was possible to predict. And, the only lesson learned was that we did not learn our lesson! Then along came Hurricane "Katrina" (August–September, 2005) and it seems we still have not learned our lesson. What happened in New Orleans was very predictable! The response, nevertheless, became a national disgrace.

Our colleague and friend Robert P. Iannone, CPP has again updated his valuable contribution on Security System Specifications, which appears in Appendix H. We are grateful, as always, to Bob for his willingness to add his experience and expertise to our work.

No work by this author would be complete without acknowledging the dean of security consulting, Phil Schiedermayer, CPP, of Lafayette, California, who passed away last year. Phil gave me my first job in security. He introduced me to Charles A. Hayden, PE, CPP, with whom I worked and served my apprenticeship as a security consultant. Charlie, a consummate professional, was a fire protection and safety engineer as well as one of the finest security professionals in the business. To have had the opportunity to work with these "giants" was truly a blessing.

A special thanks also goes to my lovely wife, Carolyn Oliver-Broder. I say a prayer of gratitude everyday for having her in my life.

We would also like to give thanks to Charles "Chuck" Sennewald, CPP, a personal and professional friend for more than 30 years. He was instrumental in helping me get my first book published. His contribution to the security profession has made him a living legend. I want to also acknowledge my business partner, Joel Villasenor, CPP, who constantly reminds me that although we may not always be perfect, we should always strive for perfection. And last, but not least, my brilliant son Roger Oliver, the family "network (ether) geek," whose assistance in electronically compiling the final chapters for this 3rd edition was invaluable.

James F. Broder, CFE, CPP, FACFE
San Marino, CA

□ □ □
□ □ □
□ □ □

Introduction

In April 2000, I was asked to make a presentation before the Research Security Administrators at their annual meeting in Northern California. The topic I chose was, "As threats evolve, so must the security profession." We believe the material presented at that conference to be even more topical today, especially with regard to what has happened to the security profession since September 11, 2001. So, with some modification, we decided to use this material as the Introduction for the third edition of this book. It is our considered opinion that some otherwise intelligent people do not fully understand what the security profession is all about. As such, many approach the issue of security with unrealistic expectations. Granted, the security profession has its strengths and weaknesses, but more important, it also has limitations. We hope to clarify some of these issues in this book. As with any good treatment of a subject, we begin by providing a dictionary definition of the word *threat,* as follows:

> An indication of impending danger or harm. A forthcoming, upcoming, possible happening event, or activity; from a security standpoint, anything that can adversely affect the assets of an enterprise or organization. The terms "threat" and "risk" are sometimes used interchangeably.

Like many professionals involved in the security profession for more than 25 years, we have seen "experts" come and go. We have also witnessed many new theories rise and fall. We have seen, especially in the heat of battle, that the truth may become obscure, or worse, lost. In fact, it has been said that "truth is the first casualty of war." We realize, in a moment of weakness, that one may be tempted to subscribe to the current fad or "security flavor of the month" theory. When this happens, fundamental principles may be ignored, and time-honored principles, as well as standards and acceptable practices, may be discarded.

We live in a society that thrives on instant gratification and the quick fix. An example of this from my own experience is the Vietnam War. After returning home from Vietnam and upon sober reflection, I realized the impracticality of some of the plans we tried to initiate in a vain attempt to pacify the rural areas of that country. Frustrated at every turn when we tried to do our duty and what we felt was right, we fell into the trap of subscribing to new and untested theories instead of sticking to time-honored principles and practices that had served us so well in the past. With the benefit of hindsight, many of us now understand that this was a terrible mistake.

As a result of years of experience as an investigator and security professional, I can now honestly identify for my clients "what does not work." Or stated differently, I may not always know what will work, but having made every mistake in the book, I know what does not work. Now, when faced with a problem that seems to defy a solution, I recall that quote from Coach Vince Lombardi, "When all else fails, go back to basics."

For those of us working in the security field, the basics are rather simple, consisting of (1) the fundamentals of the game, (2) the principles and standards of the profession, and (3) the acceptable practices of the craft, trade, or profession in which we are employed.

Having defined the subject and established the above foundation, let us examine some of the fundamentals, principles, and practices of the security game as I have come to understand them.

The Security Mission

The primary role of security is to prevent and deter. This is our mission, and it must never be confused with the missions of law enforcement or intelligence, both of which are entirely different. In an effort to explain the fundamental role of security versus that of law enforcement, I have used the concept of two large overlapping circles. One is labeled *security* and the other *law enforcement*.

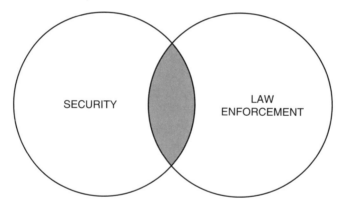

It is the overlapping area of the two circles that seems to confuse some people. The simple explanation is that sometimes security professionals make arrests, just as sometimes law enforcement professionals engage in crime prevention. In neither case, however, is that their primary mission. We used this diagram and explanation when briefing the board of directors of a large metropolitan transit district. The board was composed of the mayor, city councilmen, and county supervisors. The topic under consideration was the use of contract security officers to provide security for the transit district, as opposed to establishing their own law enforcement agency.

Before we could proceed to solve the problem, we had to ensure that everyone was operating from the same basic set of fundamentals with regard to the missions and roles played by the competing interest. One might expect that an audience of this composition would come to the table with a full understanding of these fundamentals. Such was not the case. As a result, a long, protracted debate was necessary to establish the ground rules before coming to an agreement on a "security mission statement."

What Are the Fundamentals Involved?

One of security's cardinal rules is that when security fails, for whatever reason, and a crime is committed, the emphasis (and jurisdiction) immediately shifts from security to law enforcement. For example, banks usually have state-of-the art security procedures and systems in place to prevent robberies. Notwithstanding, banks occasionally get robbed. Robbery is a crime, and the police are responsible for solving crimes.

Now this fundamental may seem simple to those of us in the security (or law enforcement) profession, but we believe the perception of the public, many of whom are well-educated executives and managers, is cloudy regarding this simple rule. As an example, every time there is a security breach or an incident at an airport (especially since 9/11), the media reports that security systems or procedures failed. And, the simple fact is that because an incident occurred that was detected by security before any harm was done, there was no failure! Is airport security perfect? No, but whoever promised us that the security installed at airports or any other location would guarantee that breaches and incidents could never occur?

We have worked with directors of loss prevention and security who tell us that their superiors hold them responsible for every security breach that results in an incident or a criminal act that occurs at their facility. Are we in the security profession perhaps taking for granted that the public, our clients, and our superiors understand the fundamentals of the security game?

Risk Assessment Is Essentially a Management Tool

This is another time-honored principle. *Risk* (for our purpose) is defined as (1) the possibility of suffering harm or loss, and (2) the danger or probability of loss occurring to one's insured assets.

Most of the risks affecting a given enterprise or organization can be readily identified and therefore predicted. Relatively few of the risks identified should worry the security professional, or cause a company to drastically alter its operational activities. A security manager who spends his time and resources defending against the less probable but more "glamorous" threats is not effectively serving his organization. The fundamental here is stick to basics. This does not mean that one's security policy and procedures should not periodically be reviewed for improvement. It does mean that time and money are valuable resources and must be allocated wisely.

Stick to the Basics

One constantly reads or sees news stories discussing the sad state of affairs regarding the "lack" of adequate security for chemical and nuclear power generating plants. There are those, selling fear and paranoia, who would have us believe that any rag-tag group of terrorists could successfully strike at these targets with impunity. The fact is that some of the best state-of-the-art security practices and procedures are found in these two industries. Another fact is that there has never been a terrorist attack against either a chemical or a nuclear power generating plant anywhere in the United States!

As an example, we were working on upgrading the security at a nuclear power generating plant located on Lake Erie, near a major metropolitan area. Plans were developed to prevent and deter any attempt at an illegal intrusion onto the plant by anyone seeking to do harm to the facility. The plant had intrusion detection alarms installed and operational at strategic locations to warn of any illegal attempt to breach the perimeter's security. This included an attempted infiltration by boat from the Lake Erie side of the facility by a determined intruder.

Shortly after the intrusion alarms became operational, we became aware of problems with the motion detectors in the intrusion alarm system. They were constantly being activated. A visual scan of the area with closed-circuit television did not, however, show the presence of an intruder, much less a group of armed terrorists. After a closer investigation, it was learned that sea gulls, flying in off Lake Erie, were landing in the "secure zone" to peck at the loose gravel, thus setting off the motion-detector alarms. This nuclear power generating plant may still be having problems with sea gulls setting off their alarms. We are, however, unaware of this or any other nuclear power generating plant having been attacked by terrorists.

Do we need security and safety at chemical and nuclear power generating plants? The answer is of course a resounding "yes." There is evidence suggesting that terrorists have made surveys of these type of facilities (among many others) as potential targets. The principles in use for securing these and other potential targets from attack are not really that complicated. They are as follows:

1. Implement sound security practices and procedures to eliminate or minimize the probability of an unfavorable event occurring.
2. Develop plans to minimize the damage and loss if an unfavorable event does occur.
3. Develop contingency and disaster recovery plans to facilitate recovery and return to safe operation at the earliest possible time.

The Nuclear Regulatory Commission (NRC) has done an outstanding job of mandating security and disaster recovery planning for nuclear facilities in the United States. The NRC also mandates periodic drills and exercises to train employees of the nuclear facilities and first responders from the surrounding communities. Contrast the NRC's record in the United States with the events in Bophal, India (1984) and Chernobyl, Russia (1986). Both events involved a breach and the absence of any means of containment, as well as the lack of contingency and disaster recovery plans to deal with such an eventuality. The result: Disaster! Both incidents have become never-to-be-forgotten examples used in every disaster recovery planning textbook on the market.

Security Is More of an Art Than a Science

This is another principle that warrants constant reminding. We have not found a single formula or application that will cover the security needs of all organizations for all situations. This is the real challenge for our profession. There are no guarantees. We are about probabilities, not guarantees! Security professionals in environments that are constantly changing are being confronted by new and unusual challenges. Granted, security systems and technologies have evolved rapidly in an effort to keep pace with these challenges. These newly developed systems are, nevertheless, only as good as the people who install, monitor, and control them. And, guess what? All systems need to be tested periodically and monitored constantly to ensure that they remain effective. But then, what is new about this concept?

Networking Works

This topic is more of a practice than a principle, but it is nonetheless important to mention. Networking among security professionals and mutual cooperation among security, law enforcement, and intelligence agencies are essential to our success in dealing with the threat of global terrorism. Together we have a wealth of knowledge unsurpassed by any other combination of professional groups in our society.

The American Society for Industrial Security (ASIS International) has more than 33,000 members worldwide. Included in this number are more than 5,000 Certified Protection Professionals (CPPs). There are more CPPs today than there were members of ASIS 30 years ago, and this society continues to grow.

Today, the Federal Bureau of Investigation (FBI) and the Central Intelligence Agency (CIA) are only too willing to cooperate with U.S. corporations here and overseas concerning industrial espionage, technology transfer, terrorism, and cyber crime. The Los Angeles Police Department and the Los Angeles Chapter of ASIS jointly sponsor a yearly seminar to discuss topics of mutual concern and to open channels of communication among professionals. The program is called *LEAPS,* for Law Enforcement and Professional Security. It is now an established forum for networking and discussing problems, solutions, and ideas. Additionally, the FBI and the CIA once had booths in the exhibit halls at the ASIS national conventions. As a result of the establishment of the U.S. Department of Homeland Security, we can now look forward to better communication and cooperation among security, law enforcement, and intelligence professionals.

The point to remember is that we in security no longer have to go it alone. No one is expected to have all the answers. Networking can usually solve any problem with which we may be confronted. In short, we have finally reached a point at which we are pooling our resources and working together for the common good. And, as one might say, "Not a moment too soon!"

As Threats Evolve, So Do We

The good news is that the security profession has always been in a constant state of evolution. At the same time, we understand that a firm belief in the use of basic principles, standards, and acceptable practices is not the same as being mired in the past. We live in a world that is constantly changing, often at a speed that is difficult to measure. But, from a professional standpoint, what is really new? Granted, security practices, procedures, and equipment that worked a few years ago need to be reviewed, tested, and upgraded for sufficiency and adequacy with regard to current operational needs. But, has this not always been the case?

Security professionals and managers must always be on a learning curve. The good ones always are! We should never allow ourselves the luxury of assuming that the countermeasures we have in place today will protect us from the dangers that may threaten us tomorrow. New situations will constantly arise to challenge our professional skills. We believe that the security profession will continue to evolve and face these new challenges, using the same fundamentals, acceptable practices, principles, and standards that have served us so well in the past.

I sincerely hope that the material presented in the following chapters will be of some help to the readers in accomplishing their security mission.

James F. Broder, CFE, CPP, FACFE
San Marino, CA

The Treatment and Analysis of Risk

1

Risk

> *"Security is more art than science. Few formulas will cover all organizations, situations and needs, and that's the beauty and challenge of our profession. We are about probabilities."*
> —Richard D. Sem, CPP, Are These Truths Self-Evident? *Security Management,* March 1998

WHAT IS RISK?

Risk is associated with virtually every activity one can imagine, but for the purpose of this text, we limit the meaning of the word *risk* to the uncertainty of financial loss, the variations between actual and expected results, or the probability that a loss has occurred or will occur. In the insurance industry, risk is also used to mean "the thing insured"—for example, the XYZ Company is the risk. Risk is also the possible occurrence of an undesirable event.

Risk should not be confused with *perils,* which are the causes of risk—such things as fire, flood, and earthquake. Nor should risk be confused with a *hazard,* which is a contributing factor to a peril. Almost anything can be a hazard—a loaded gun, a bottle of caustic acid, a bunch of oily rags, or a warehouse used for storing highly flammable products, for example. The end result of risk is loss or a decrease in value.

Risks are generally classified as "speculative" (the difference between loss and gain—for example, the risk in gambling) and "pure risk," a loss/no-loss situation, to which generally applies insurance.

For the purposes of this text, the divisions of risk are limited to three common categories:

- Personal (people assets)
- Property (material assets)
- Liability (legal issues) that could affect both of the above. Here we include such problems as errors and omissions, wrongful discharge, workplace violence, sexual harassment, and last but not least, what has become the biggest legal issue to plague the business community, third-party liability. This latter issue has given rise to a whole new profession, that of the "security expert witness." The subject is so important to the security professional that we have devoted a chapter to it called Crime Prediction (see Chapter 10).

WHAT IS RISK ANALYSIS?

Risk analysis is a management tool, the standards for which are determined by whatever management decides it wants to accept in terms of actual loss. To proceed in a logical manner to perform a risk analysis, it is first necessary to accomplish some basic tasks:

- Identify the assets in need of protection (people, money, manufactured products, and industrial processes, to name a few).
- Identify the kinds of risks (or perils) that may affect the assets involved (kidnapping, extortion, internal theft, external theft, fire, and earthquake, for example).
- Determine the probability of risk occurrence. Here one must keep in mind that the task of making such a determination is not a science, but an art—the art of projecting probabilities. Remember this rule: "Nothing or no one can ever be made 100 percent secure."
- Determine the impact or effect on the organization in dollar values when possible, if a given loss does occur.

These subjects are discussed further in later chapters of this text.

RISK ASSESSMENT

Risk assessment analysis is a rational and orderly approach, as well as a comprehensive solution, to problem identification and probability determination. It is also a method for estimating the expected loss from the occurrence of an adverse event. The key word here is *estimating* because risk analysis will never be an exact science—we are discussing probabilities. Nevertheless, the answer to most, if not all, questions regarding one's security exposures can be determined by a detailed risk assessment analysis.

What Can Risk Analysis Do for Management?

Risk analysis provides management with information on which to base decisions. A thorough risk analysis can provide answers to many questions, such as, Is it always best to prevent the occurrence of a situation? Is it always possible? Should the policy also consider how to contain the effect a hazardous situation may have? (This is what nuclear power plants prepare for.) Is it sufficient simply to recognize that an adverse potential exists and to do nothing for now but be aware of the hazard? (The analogy is being self-insured.) The eventual goal of risk analysis is to strike an economic balance between the impact of risk on the enterprise and the cost of implementing prevention and protective measures.

A properly performed risk analysis can have many benefits, a few of which are as follows:

- The analysis will show the current security posture (profile) of the organization.
- It will highlight areas in which greater (or lesser) security is needed.

- It will help to assemble some of the facts needed for the development and justification of cost-effective countermeasures (safeguards).
- It will increase security awareness by assessing and then reporting the strengths and weaknesses of security to all organizational levels from management to operations.

Risk analysis is not a task to be accomplished once and for all; it must be performed periodically if one is to stay abreast of changes in mission, facilities, and equipment. Also, because security measures designed at the inception of a program for building or expansion generally prove to be more effective than those superimposed later, risk analysis should have a place in the design or building phase of every new facility. This is being done increasingly as architects have begun to realize the importance of planning for security.

The major resource required for performing a risk analysis is trained manpower. For this reason, the first analysis is the most expensive. Subsequent analyses can be based in part on previous work history, and the time required to do a subsequent survey should decrease to some extent as experience and empirical knowledge are gained.

The time allocated to accomplish the risk analysis should be compatible with its objectives. Large facilities with complex, multishift operations and many files of historical data require more time to gather and review the necessary data than does a single-shift, limited-production location. If meaningful results are to be achieved, management must be willing to commit the resources necessary to accomplish the mission. It is best to delay or even abandon a security survey unless and until the necessary resources are made available to complete it properly. In this regard, the security professional should be ever mindful of the legal risk associated with being "on notice." After a situation is identified as being a "security risk" and is brought to management's attention, corrective action must be taken. Not to take this action would expose the company to legal liability should an otherwise preventable incident occur.

Role of Management in Risk Analysis

The success of any risk analysis undertaking is strongly contingent on the role top management plays in the project. Management must support the project and express this support to all levels of the organization. Management must delineate the purpose and define the scope of risk analysis. It must select a qualified team and formally delegate the authority necessary to accomplish the mission. Finally, management must review the team's findings, decide which recommendations need to be implemented, and then decide on and establish the order of priorities for implementing the recommendations made in the survey report.

Personnel who are not directly involved in the survey and the analysis process must be prepared to provide information and assistance to those who are. In addition, all employees must abide by any inconvenience and the possible temporary limitations of their activity that may result from actions of the survey team. Management should make it clear to all employees that it intends to rely on the final product and base its security decisions on the findings of the risk analysis team's report. The scope of the project should be well defined, and the statement of scope should spell out the parameters and depth of the analysis. It is often equally important to state specifically and in writing what the survey is *not* designed to accomplish or cover; this eliminates any misunderstandings at the start of the project. An example might be the exclusion from a security survey of safety and evacuation procedures in a hospital setting.

At this point, it may be helpful to define and explain two other terms that are sometimes used interchangeably with risk: *threat*—anything that could adversely affect the enterprise or the assets; and *vulnerability*—a weakness or flaw, such as holes in a fence, or virtually anything that may conceivably be exploited by a threat. Threats are most easily identified and organized by placing each in one of three classifications or categories: natural hazards (floods), accidents (chemical spills), or intentional acts (domestic or international terrorism). Vulnerabilities are most easily identified by interviewing long-term employees, supervisors, and managers in the facility; by field observation and inspection; and by reviewing incident reports. In the case of security hardware and electronics, tests can be designed and conducted to highlight vulnerabilities and expose weaknesses or flaws in the operations of the system. Examples would be an out-of-date access control system or inadequate procedures to ensure an orderly evacuation of a high-rise office building in an emergency.

Threat occurrence rates and probabilities are best developed from reports of occurrences or incident reports, whenever these historical data exist. If reports containing this information do not exist, it may be necessary to develop the information from other sources. This can be accomplished by conducting interviews with knowledgeable people or projecting data based on an educated guess, supported by studies in like industries and different locations.

RISK EXPOSURE ASSESSMENT

Before any corrective action can be considered, it is necessary to make a thorough assessment of one's identifiable risk exposure. To accomplish this, it is essential that three factors be identified and evaluated in quantitative terms. The first is to determine (identify) the types of loss or risk (perils) that can affect the assets involved. Here examples would be fire, flood, burglary, robbery, and kidnapping. If one of these were to occur (for now we will consider only single, not multiple, occurrences), what effect would the resulting disruption of operations have on the company? For example, if vital documents were destroyed by fire or flood, what would the effect be on the ability of the company to continue operating? There is a saying common to protection professionals: "One may well survive a burglary, but a major fire can put you out of business forever." If the chief executive officer, on an overseas trip, were to be kidnapped by a terrorist group (or even suffer a serious heart attack), who would make the day-to-day operating decisions in his or her absence? What about unauthorized disclosure of trade secrets or other proprietary data? After all the risk exposures are identified (or as many as possible), one must proceed to evaluate those identified threats that, should they occur, would produce losses in quantitative terms—fire, power failure, flood, earthquake, and unethical or dishonest employees, to name a few of the more common risks necessary of consideration.

To do this, we proceed to the second factor: estimate the probability of occurrence. What are the chances that the identified risks may become actual events? For some risks, estimating probabilities can be fairly easy. This is especially true when we have documented historical data on identifiable problems. For example, how many internal and external theft cases have been investigated during the past year? Other risks are more difficult to predict. Workplace violence, embezzlement, industrial espionage, kidnapping, and civil disorder may have never occurred or may have occurred only once.

The third factor is quantifying (prioritizing) loss potential. This is measuring the impact or severity of the risk, if in fact a loss does occur or the risk becomes an actual event. This exercise is not complete until one develops dollar values for the assets previously identified. This part of the survey is necessary to set the stage for classification, evaluation, and analysis, and for the comparisons necessary to the establishment of countermeasure (safeguard) priorities. Some events or kinds of risk with which business and industry are most commonly concerned are as follows:

- Natural catastrophe (tornado, hurricane, seismic activity)
- Industrial disaster (explosion, chemical spill, structural collapse, fire)
- Civil disturbance (sabotage, labor violence, bomb threats)
- International and domestic terrorism
- Criminality (robbery, burglary, pilferage, embezzlement, fraud, industrial espionage, internal theft, hijacking)
- Conflict of interest (kickbacks, trading on inside information, commercial bribery, other unethical business practices)
- Nuclear accident (for example, Three Mile Island, Detroit Edison's Enrico Fermi #1)

Some of these events (risks) have a low, or zero, probability of occurrence; also, some are less critical to an enterprise or community than others even if they do occur (fire versus burglary, for instance). Nevertheless, all of the identified possible events could occur and are thus deserving of consideration.

Examples include the nuclear accident at Chernobyl in the Soviet Union in 1987 and the chemical gas disaster ("breach of containment") at the Union Carbide plant in Bhopal, India in 1984. Also, there are today in the United States chemical and nerve gas weapons stored in bunkers at military depots near populated areas. Do contingency plans exist to deal with these risks in the event of accidental fire, leak, or explosion? Are disaster drills and exercises conducted periodically to test the effectiveness of the contingency plans, if they exist? There are contingency plans for breach of containment and other industrial accidents in nuclear power generating plants in the United States, and these are strictly enforced by the Nuclear Regulatory Commission (NRC), which requires periodic drills to rehearse the plans.

In the following chapters, we continue to discuss vulnerabilities and threat identification, as well as risk measurement and quantification.

2 ▢▢▢
   ▢▢▢
   ▢▢▢

Vulnerability and Threat Identification

"Before the question of security can be addressed, it is first necessary to identify those harmful events which may befall any given enterprise."
—Charles A. Sennewald, CPP, Security Consultant, Author, and Lecturer

RISK IDENTIFICATION

In systems security, the primary purpose of vulnerability identification or threat (exposure) determination is to make the task of risk analysis more manageable by establishing a base from which to proceed. When the risks associated with the various systems and subsystems within a given enterprise are known, the allocation of countermeasures (resources) can be more carefully planned. The need for such planning rests on the premise that security resources, like all other resources, are limited and therefore must be allocated wisely.

Risk control begins, logically, with the identification and classification of risk. To accomplish this task, it is necessary to examine or survey all the activities and relationships of the enterprise in question and to develop answers to these basic considerations:

- Assets—What does the company own, operate, lease, control, have custody of or responsibility for, buy, sell, service, design, produce, manufacture, test, analyze, or maintain?
- Exposure—What is the company exposed to that could cause or contribute to damage, theft, or loss of property or other company assets, or that could cause or contribute to personal injury of company employees or others?
- Losses—What empirical evidence is available to establish the frequency, magnitude, and range of past losses experienced by this and other companies located nearby, performing a like service, or manufacturing the same or similar products?

Obviously, the answers to these questions and any additional questions that may be raised when conducting initial inquiries will be the basis for the identification of risks and eventually the evaluation of risks that may have a negative effect on the enterprise in question.

Security professionals use many techniques to develop data for risk identification. They may review company policies, procedures (or take into account their absence), the structure of the organization, and its activities to ascertain what risks have been identified and to what extent they are perceived as management responsibilities. They may review insurance and risk-related files, including claims and loss records. Interviews with the heads of departments that have experienced loss exposures can develop vital information on the organization and functioning of loss-control procedures, if in fact any exist. Conducting observation tours and inspections and interviewing management and other personnel in enough locations and activities will help develop a comprehensive picture of the company's risk exposures as a basis for later evaluation of existing loss control procedures and their effectiveness.

The tools necessary to accomplish the above tasks are the ability to conduct thorough interviews; the ability to conduct inspections and field observations of operations, procedures, manpower utilization, hardware, and electronics used in security systems; and the ability to identify, obtain, and analyze pertinent records.

Another technique one may use is to develop asset data. To do this, one needs to completely identify all company assets, tangible and intangible, in terms of quantity and quality. One then locates all company assets and identifies obvious exposures that may exist at these locations. Next, one must determine the value of these assets in terms of actual dollars. This should be broken into the following three categories:

- Owned assets $_____
- Leased assets $_____
- Facility losses $_____
  Total tangible assets $_____
  Total intangible assets $_____
  Grand total $_____

The identification of all company assets, coupled with a history of loss exposure for the company and other companies similarly located and engaged, will normally be sufficient to identify most of the major risks involved.

After this identification procedure, the security survey or inspection can be limited to those risks or exposures that specifically relate to the enterprise in question. These risks will usually include most, if not all, of the following:

- Crime losses, such as burglary, theft (internal and external), fraud, embezzlement, vandalism, arson, computer abuse, bomb threat, theft of trade secrets and industrial espionage, forgery, product forgery and trademark infringement, robbery, extortion, and kidnapping—to name the most common crime risks encountered by business and industry. Some others are:
- Cargo pilferage, theft, and damage
- Emergency and disaster planning
- Liability of officers and directors
- Environmental controls as directed by occupational safety and health codes
- Damage to property from fire, flood, earthquake, windstorm, explosion, building collapse, falling aircraft, and hazardous processes

- Comprehensive general liability arising from damage caused by any activity for which the entity can be held legally liable
- Business interruption and extra expense. An evaluation of this type of risk may require a detailed study of interdependencies connecting various segments of the entity and outside suppliers of goods and services. The study may indicate the need for extensive disaster recovery planning, for the reduction of risks not previously identified.
- Errors and omissions liability
- Professional liability
- Product liability

As can be seen from even a cursory review of these risks, the scope of risk identification alone, separate from risk evaluation and risk control, presupposes a degree of education and practical knowledge not often possessed by the average security manager. This implies that the person who is charged with this responsibility should have the education, training, and practical experience necessary to seek out, recognize, and thus identify not only the risk involved but also its applicability to the enterprise in question. The process of risk identification, evaluation, and control in any dynamic organization, public or private, requires constant attention by professionals who possess the necessary knowledge and tools to accomplish these tasks. This knowledge is best acquired by study and practical experience. There is, however, no substitute for "hands on" experience when it comes to the analysis of risks.

EXAMPLES OF THE PROBLEMS OF IDENTIFICATION

I was given the assignment of conducting a security survey for a chain of fast-food restaurants. At the initial meeting with the management staff of this chain, reports from the company's internal audit division were furnished for our review. These reports showed all crime-related losses for a 12-month period. They contained statistics that showed a disturbingly high incident rate for "robbery." One of the firm's top management people stated that he had a growing concern that a part-time high school cashier working in one of their restaurants might get shot in the course of a robbery because of the absence of procedures instructing employees how to deal with a robbery.

Field inspections of a representative number of these restaurants produced evidence that the crime problem most often reported was burglary (breaking and entering), followed closely by internal theft. "Armed robbery," or just plain robbery, as perceived by management, although admittedly always a dangerous situation, was not as frequent or as serious a problem as management had been led to believe by the statistics in the audit reports. Further inquiry revealed that the terms *robbery* and *burglary* were being used indiscriminately and, in many cases, synonymously. It was only after identifying the real problem that we could proceed to develop the proper procedures and allocate the necessary resources to address and then solve management's concern.

In another case, I met with the management of a national corporation that, among other things, printed negotiable instruments. The purpose of the meeting was to develop a mutual agreement regarding the scope of a security survey to be conducted at one of its West Coast plants. At the outset, one of the management representatives asked, "Have you ever conducted a security survey at a plant that prints negotiable instruments?" The simple fact at the time was that I had not. Nevertheless, the answer I gave was, "No, I haven't, but that really doesn't matter. It is immaterial to me if your plant manufactures widgets or prints negotiable instruments. You will either have a security program or you will not. If you do, it will either be functional, or it will not. In either event, it can be evaluated, and we can determine if the state of the security in existence at the plant is adequate, given the unique requirements for protection that this type of production requires. If we feel the system is not functional, we will make recommendations to upgrade the quality and the quantity of the security safeguards necessary to accomplish this goal. If you have no program for security, we will design one for your consideration to meet the above requirements."

In security, as with many other disciplines, we deal in acceptable practices and principles. These remain fairly constant, regardless of the product involved or the environment encountered. The end result is loss control. This means that one either has or does not have an adequate security system. One way to find out is to conduct a survey and identify those harmful events that may interfere with the company's objectives, as they are defined by the management of the enterprise in question.

Security Checklist

Until now we have been discussing some of the techniques and tools that the security professional needs to develop data for risk identification. Very often, security checklists are used to facilitate the gathering of pertinent information. These checklists take many forms. They can be simple lists of yes-or-no questions, or open-ended questions requiring narrative responses. They may be brief and narrowly focused on the specific operation or activity in question, or they may be broader in scope and cover security concerns common to all the company's operations. No matter what its appearance, the purpose of a security checklist is to provide a logical recording of information and to ensure that no important question goes unasked.

The checklist is usually the backbone of the security survey or audit. This subject is covered extensively in Chapters 7 and 9.

The following is a general or introductory security survey checklist. It was designed for company-wide use; therefore, it may include many items that are appropriate in some situations but not in others. Additional checklists can be found in the appendices of the book.

    I. Policy and Program
      1. Top management established a security policy?
        a. Policy published?
        b. Part of all managers' responsibility?
        c. Designated individual to establish and supervise security program?
      2. Top manager accessible to security supervisor?
      3. Any regulations published? (Attach copy)
      4. Disciplinary procedures?
        a. In writing?
        b. Specify offenses and penalties.

    c. Incidents recorded?

    d. Review by management?

    e. Uniformly enforced?

  5. Any policy on criminal prosecution?

    a. Number of prosecutions attempted during past 5 years?

    b. Number of convictions?

II. Organization

  1. Security supervisor full time?

    a. If part time, percentage of time spent on security?

    b. Describe chain of command from security supervisor to plant manager.

  2. Number of full-time security personnel?

  3. Number of personnel performing security duties each shift?

    a. Do they perform nonsecurity duties concurrently?

    b. Do security duties have first priority?

  4. Have security personnel received security training?

  5. Are written reports made of incidents?

  6. Is there follow-up investigation of incidents?

  7. Background investigation of security personnel?

  8. Guards?

    a. Number?

    b. Proprietary or contract service?

    c. If contracted service, does plant security supervisor interview and select?

    d. Is there a written contract for guard service?

      i. Management's terms and conditions included?

    e. Written guard orders? (Attach copy)

    f. Weapons carried? (List type—for example, pistols, mace)

      i. If yes, who inspects?

      ii. Company furnished?

    g. Make electronically recorded tours?

    h. Guards make written tour reports?

    i. Frequency of tours?

    j. Tour pattern varied?

    k. Number of report stations?

    l. Guards submit written report each shift? (Attach copy of form)

    m. Have guards received any formal training? (Describe on reverse side)

    n. Appearance of guards and uniforms?

  9. Procedures

    a. Have security procedures been published? (Attach copy)

    b. Distributed to all those affected?

    c. Revised when conditions change?

    d. Used to conduct periodic audits and drills?

  10. Does the security supervisor maintain contact with local law enforcement agencies to keep abreast of criminal activities and potential disorder in the community?

III. Control of Entry and Movement
   1. Is identification required of all persons entering?
   2. Are there periodic 100 percent checks of identification?
   3. How often? By whom?
   4. Are all visitors registered?
   5. How are employees distinguishable from visitors? (Explain on reverse)
   6. Are all visitors escorted at all times?
   7. Is there control of employee movement between areas within the plant? (Describe on reverse)
   8. Are supervisors instructed to challenge strangers in their work areas?
      a. Do they? Nearly always? Sometimes? Never?
   9. Are periodic traffic counts made at all points of entry as a means of detecting need for schedule change?
   10. Are identification badges issued to all employees? Wearing enforced?
IV. Barriers (Fences, Gates, Walls, etc.)
   1. Is there a continuous barrier around the entire plant property?
      a. A major portion of it?
      b. Areas outside barrier? (List)
   2. Fencing
      a. Eight feet high?
      b. Two-inch-square mesh?
      c. Eleven-gauge or heavier wire?
      d. Topped by three strands barbed wire or selvage?
      e. In good repair?
      f. Within 2 inches of firm ground at all points?
      g. Securely fastened to rigidly set posts?
      h. Metal posts set in concrete?
      i. Where attached to buildings, gaps not more than 4 inches?
      j. Gates in good repair? (How many?)
         i. Gates same height and construction as fence?
         ii. Open only when required for operations?
         iii. Locked other times?
         iv. Equipped with alarm? (How many?)
         v. Guarded when open?
         vi. Under surveillance when open? How?
      k. At least 10 feet of clear space both sides of fence?
   3. Along embankments, fence on top or 20 feet from bottom?
   4. Walls as perimeter barriers
      a. At least 8 feet high?
      b. All doors equipped with alarm device or under surveillance?
      c. Means of surveillance?
      d. Windows:
         i. Permanently closed?
         ii. Accessible for removal of property?
         iii. Can be used for entry or exit?

        iv. Protected by bars or heavy screen?

        v. Equipped with alarm?

    e. Final exit (perimeter) doors:

        i. Guarded?

        ii. Alarm equipped? What type?

        iii. Controlled by security personnel? Controlled by devices?

        iv. Strong enough to resist heavy impact?

        v. Hinge pins concealed from outside or security type?

V. Lighting

1. Entire perimeter lighted?
2. Strip of light on both sides of fence?
3. Illumination sufficient to detect human movement easily at 100 yards?
4. Lights checked for operation daily before darkness?
5. Extra lighting at entry points and points of possible intrusion?
6. Lighting repairs made promptly?
7. Is the power supply for lights easily accessible (for tampering)?
8. Are lighting circuit drawings available to facilitate quick repairs?
9. Switches and controls
   a. Protected?
   b. Weatherproof and tamper resistant?
   c. Accessible to security personnel?
   d. Inaccessible from outside the perimeter barrier?
   e. Master switches centrally located?
10. Good illumination for guards on all routes inside the perimeter?
11. Materials and equipment in receiving, shipping, and storage areas adequately lighted?
12. Bodies of water on perimeter adequately lighted?
13. Auxiliary source of power for protective lighting?

VI. Locks and Keys

1. Responsibility for control of locks and keys assigned to security supervisor?
2. Does he or she control locks and keys to all buildings?
3. Does he or she have overall authority and responsibility for issues, changes, and replacements?
4. Plant manager approve formula for issuing keys?
5. Managers approve issue of keys to their area?
6. Any keys issued to nonemployees?
7. Recipient sign receipt for key?
   a. Receipt show building and room number, date, name of authorizing manager?
   b. Receipt acknowledge obligation to turn in, report loss, not to duplicate?
8. Keys issued solely because of operational need for recipient to have key?
9. Lock and key control procedures and regulations in writing? (Attach copy)

10. All keys recovered from terminating employees?
11. Master keys not marked as such?
12. Spare keys stored under double lock or in combination-locked, fireproof cabinet?
13. Access to spare keys restricted to security supervisor and one other manager?
14. Locks changed immediately upon theft or loss of keys?
15. Locks on perimeter doors and gates changed annually?
16. Padlocks changed or rotated annually?
17. Manufacturer's serial number on padlocks obliterated and replaced by plant code number?
18. Padlock locked to hasp or staple when door or gate is open (to prevent substitution)?
19. Locks on inactive doors and gates checked regularly for evidence of tampering?
20. Door locks installed so that bolt extends ½ inch into jamb?
    a. Bolt covered by steel cover plate between door and jamb to prevent levering?
21. Combination locks
    a. Combination changed:
        i. Annually?
        ii. When unauthorized person may have learned?
        iii. When knowledgeable person leaves or transfers?
    b. Combination memorized? (Not written!)
    c. Combination numbers: one odd, one even, one divisible by five? (Any sequence OK)
    d. Combination disclosed on basis of operational necessity (not convenience)?
22. Perimeter doors
    a. Lock installed without keyway in outside knob?
    b. Deadbolt locks in doors that must be unlocked from outside?
23. Safes
    a. Of substantial construction?
    b. Rated (labeled) for fire resistance?
    c. Rated (labeled) for burglary resistance?
    d. Lighted at night?
    e. Covered by proximity or motion detection alarm?

VII. Alarms
    1. Fire alarms
        a. Water flow? Is water pressure present?
        b. Valve condition?
        c. Water temperature?
        d. Particles of combustion detection?
        e. Heat or smoke sensing?
        f. Monitored continuously by:
            i. Contract central station?
            ii. Proprietary station?

           iii. Direct connection to police or fire department?

      g. Tested regularly and tests recorded?

      h. Additional functions performed by alarm system (for example, shuts off computer power, lights, heating, air conditioning)?

  2. Intrusion alarms

      a. Protects all of plant perimeter?

      b. Protects high-value storage areas?

      c. Protects other internal areas? (List)

      d. Types of sensors? (List)

      e. Proprietary or central station supervision?

      f. Regular recorded tests? How often?

  3. Closed circuit television

      a. Used for surveillance only?

      b. Used for access control?

      c. Monitored continuously?

VIII. Communications

  1. Separate communications for security and emergency use?

      a. Telephone?

      b. Radio, pagers, cell phones?

  2. If radio shared with other users, can security override?

  3. Is there a means of contacting guard on patrol immediately? How?

  4. Procedure for contacting local police and fire departments?

  5. Means of alerting employees to emergency? How?

IX. Property Control (Equipment, Material, Tools, Personal Property)

  1. Covered by written procedures?

  2. Specified form?

      a. Serial numbered?

      b. Multipart to provide separate audit trails?

  3. Signed authorization required? Approved by higher level than beneficiary?

  4. All transactions monitored at exit?

  5. All exits controlled?

  6. All transactions audited by third party (other than security)?

  7. Follow-up on late returns (of borrowed items)?

  8. Control points between work area and parking area?

  9. Spot-checks of trucks and other vehicles?

 10. All company tools (except small hand tools) marked with permanent company identification?

 11. Employees sign receipt for tools and equipment issued?

 12. Tools and desirable items secured in locked cages or rooms?

      a. Inventoried frequently?

 13. All losses reported?

      a. Follow-up investigation?

      b. Written record?

      c. Statistics compiled? Reported to top management?

 14. Shipping, receiving, and storage

      a. Guarded or within protected area?

b. Under security of supervisor surveillance?
c. Continuous spot-checks of complete shipments and receipts versus documents by other than shipping/receiving clerks?
d. Outside drivers permitted inside plant?
e. Vehicles locked?
f. Tailgate check of existing vehicles?
g. Storage areas under separate lock control when unattended?
h. All withdrawals from stock recorded? Records provide separate audit trails?
i. Continuous spot-checks of waste containers?

15. Scrap and salvage
a. Written procedure for collection and disposal of scrap and salvage?
b. Sealed bids required?
c. Disposal action documented?
d. Purchaser selected by management (other than administering employee)?
e. Estimated annual sales of scrap and salvage?
f. Any sold or given to employees? (Explain on reverse or attach procedure)
g. Stored in a locked secure area?
h. Waste spot-checked for saleable scrap and salvage?
i. Classified and separated as to value?
j. Spot-checked for "high grading"?
k. Removed from premises under:
    i. Signed authorization on specified form?
    ii. Surveillance of third-party employee, for example, accounting or security?
    iii. Verifies class and quantity?
    iv. Compares with purchaser's receipt?
l. Auditors review all transactions?
m. Are quantities within normal limits for this type of operation?

X. Emergency Planning
1. Plans for reaction to imminent or actual:
a. Fire?
b. Explosion?
c. Flood or tidal wave?
d. Hurricane?
e. Earthquake?
f. Disorder?
g. Aircraft accident?
h. Bomb threats and bombs?
2. Responsibilities spelled out?
3. Responsible individuals designated?
4. Organization(s) completely staffed?
5. Periodic rehearsals of:
a. All personnel?

        b. Key personnel?
- 6. Have critical features of plant and equipment been identified?
  - a. Protected by barriers, access control, and lighting?
- 7. Coordinated with local public safety and disaster organizations?
- 8. Include plans for post-disaster recovery?
- 9. Identify resources available and required?

XI. Personnel Screening
1. Written, signed employment application required? Omissions not tolerated?
2. All candidates interviewed?
3. Investigation to verify?
  - a. Previous employment
    - i. Employers?
    - ii. Dates?
    - iii. Position and duties?
    - iv. Salary?
    - v. Quality of performance?
  - b. Education?
  - c. Criminal record?
  - d. Reputation?
  - e. Medical record
    - i. Illnesses?
    - ii. Physical handicaps and limitations?
    - iii. Work injuries?
    - iv. Occupational illnesses?
4. Special screening of candidates for fiduciary positions? Describe.

XII. Comments

Add any information you consider useful in arriving at a realistic assessment of the security in the plant. After all the significant loss potentials are identified, the next step is to evaluate the identified threats or vulnerabilities that may affect the enterprise and thus conceivably produce losses. The completed survey forms should be retained in the audit file for later use in the preparation of the final survey report.

3 ⬜⬜⬜
⬜⬜⬜
⬜⬜⬜

Risk Measurement

"Risk can not always be eliminated. Properly identified, however, it can usually be managed."
—*Risk Assessment Guidelines, General Security*, ASIS International, 2003

Risk measurement (quantification) is an essential element for later use in determining the impact (cost) of an unfavorable event on the enterprise. It can also aid in predicting how often such an event may occur in a given period of time. Two necessities for performing risk measurement and quantification are a quantitative means of expressing potential cost and a logical expression of frequency of occurrence. Both must admit low as well as high frequencies of event occurrence.

There is no better way to state the impact of an adverse circumstance—whether the damage or cost is actual or abstract, or the victim a person, a piece of machinery, or the entire facility—than to assign it a monetary value. Ascertaining the cost of any adverse event is the logical way to equate value in our society. For a company that is concerned with cost (and which are not?), it is the only way. Because budgets and other financial matters are normally organized on a yearly basis, a year is obviously the most suitable time period to use in expressing frequency of occurrence of threats. Of course, some threats may occur only once in a period of years, such as the 100-year flood. Others may occur daily or many times a day, such as internal theft. Each, however, can be measured in dollars as well as frequency of occurrence.

COST VALUATION AND FREQUENCY OF OCCURRENCE

It is much more difficult to say that something happens every 1/73 of a year than that it happens, say, five times a day. It is also inconvenient to work with such fractions. For this reason, the transmutation of 1,000 days to 3 years, as shown below, has evolved. This method avoids unwieldy fractions yet maintains the flexibility of working with high-probability events in days and low-probability events in years.

In most cases, it is neither necessary nor desirable to make precise statements of impact and probability. The time needed for the analysis will be considerably reduced, and its usefulness will not be decreased, if impact (i) and frequency (f) correlations are given in factors of 10. It does not really matter to the overall estimation of threats whether the cost of the threat is valued at $110,000 or $130,000, or whether the anticipated frequency is 8 or 12 times a year. If at the time of deciding upon safeguards it becomes necessary to refine specific items, then by all means do so and end the argument! What is essential in the beginning is simplifying the measurement and quantification process, for reasons of efficiency and speed. This will facilitate the task by decreasing the amount of time spent on the analysis.

If the cost valuation (impact) of the event is:

$10, let i = 1
$100, let i = 2
$1,000, let i = 3
$10,000, let i = 4
$100,000, let i = 5
$1,000,000, let i = 6
$10,000,000, let i = 7
$100,000,000, let i = 8.

If the estimated frequency of occurrence is:

Once in 300 years, let f = 1
Once in 30 years, let f = 2
Once in 3 years, let f = 3
Once in 100 days, let f = 4
Once in 10 days, let f = 5
Once per day, let f = 6
Ten times per day, let f = 7
One hundred times per day, let f = 8.

Annual loss expectancy (ALE) is the product of impact and frequency. When using the values of f and i derived from the conversion tables (listed above), you can approximate the value of ALE by the following formula:

$$ALE = 10^{(f+i-3)}/3$$

No weighting factors have been introduced into the formula; the change is only for the purpose of accommodating the converted values. An even faster way to determine ALE is to use the matrix shown in Table 3.1, or alternatively, to develop one's own matrix. It would be impossible to list all the undesirable events that could plague any given security project. Most projects involving facilities, whether high technology, refinery, manufacturing, or service, have more things in common than one would suspect at first glance (for example, fires on a cruise liner versus terrorism on the high seas). These common-origin problems have to be spliced or woven into the matrix along with events or occurrences peculiar to the particular analysis at hand.

A thorough understanding of the elements that may affect frequency estimation is one of the keys to risk measurement. The following are some common elements that

Table 3.1   Determination of Annual Loss Expectancy = ALE

	Value of f							
	1	2	3	4	5	6	7	8
1					$300	$3K	$30K	$300K
2				$300	3K	30K	300K	3M
3			$300	3K	30K	300K	3M	30M
4		$300	3K	30K	300K	3M	30M	300M
5	$300	3K	30K	300K	3M	30M	300M	
6	3K	30K	300K	3M	30M	300M		
7	30K	300K	3M	30M	300M			

(Value of i labels the rows 1–7)

deserve consideration:

- Access—Is access difficult, limited, or open? Can an intruder gain access easily, or is it difficult? Can any employee do the same? What are the access criteria?
- Natural disasters—What kinds of natural disasters might realistically occur? To what degree would damage occur? How would it affect processing, storage, or supplies? How would loss of power or other utilities affect the entity?
- Environmental hazards—What special hazards are inherent in the operation? What is nearby? Are there any explosives, gasoline, or flammable objects in the area? Unused buildings next door? What can be the aftermath of fire? Water damage? Loss of stock, material? Proximity of fire and police departments?
- Facility housing—What protective devices are installed, or can be installed? Anti-intrusion alarm systems, electronic access control systems? How is the building constructed? Type of roof? Sprinklers? What kind of flooring? What is flammable?
- Work environment—What is the relationship between personnel and management? (Loyal? Suspicious?) What are the aggravations of employees? Past labor history? How well do supervisors know employees? What is management's attitude toward employee dishonesty? (Condone? OK within bounds? Dismissal?) How open are lines of communication between employees and supervisors? Supervisors and upper management?
- Value—How much can an intruder profit? How much damage could result in the worst-case scenario? How much can a dishonest employee gain? How long before an intrusion will be detected? What is security response capability, time?

PRINCIPLES OF PROBABILITY

At this point, some statements about the nature of risk must be expounded. What we have stated so far is a simple approach to identifying and measuring risk. *Risk* is the possible happening of an undesirable event. An *event* is something that can occur, a definable occurrence. When the event happens, it can be described. Security counter-measures are designed to protect against harmful events. For this reason, the question, "Is a system secure?" is meaningless. What should be asked is, "Is the system protected against events believed to be harmful"?

Any event can be described in at least two ways: it may be described in terms of the damage it will present if it occurs; or, it may be considered in terms of the probability of its occurrence. A risk, however, should be described in terms of its possibility of occurrence and its capacity for potential loss.

The study of the possibility of occurrence is known as *probability*. The principles that follow are based on philosophical (rather than mathematical) proofs derived in 1792 by the Marquis de Laplace in his *Théorie Analytique des Probabilités*. Excerpts from this classical treatise are reprinted below, in part. Laplace established 10 principles of probability, as quoted below.

1. Probability is defined as the ratio of the number of favorable cases to all possible cases.
2. If the cases are not equally possible then the probability is the sum of the possibilities of each favorable case.
3. When the events are independent of each other, the probability of their simultaneous occurrence is the product of their separate probabilities.
4. If two events are dependent on each other, then the probability of the combined event is the product of the probability of the occurrence of the first event and the probability that the second event will occur given the occurrence of the first event.
5. If the probability of a combined event first phase and that of the second phase is determined, then the second probability divided by the first is the probability of the expected event drawn from an observed event.
6. When an observed event is linked to a cause, the probability of the existence of the cause is the probability of the event resulting from the cause divided by the sum of the probabilities of all causes.
7. The probability that the possibility of an event falls within given limits is the sum of the fractions [#6 above] falling within these limits.
8. The definition of mathematical hope is the product of the potential gain and the probability of obtaining it.
9. In a series of probable events, of which some produce a benefit and the others a loss, we shall have the advantage that results from it by making a sum of the products of the probability of each favorable event by the benefit that it procures, and subtracting from this sum that of the products of the probability of each unfavorable event by the loss that is attached to it. If the second sum is greater than the first, the benefit becomes a loss and hope is changed to fear.
10. Moral hope is defined as the relation between its absolute value divided by the total assets of the involved entity. This principle deals with the relation of potential gain to potential loss and describes the basis for not exposing all assets to the same risk.

Readers and students of this text have asked me to explain Laplace's theory in words of one syllable. Unfortunately, I am not astute enough to do so. The theory is merely set forth here for those readers who may desire a more precise methodology by which to arrive at probability in their unique environment or studies. For our purposes, however, the simpler the application, the better. Excessive refinement of the risk measurement process, in our opinion, would only serve to further delay the outcome of the project, without adding much benefit to the ultimate solution.

PROBABILITY, RISK, AND SECURITY

When security is defined as the implementation of a set of acceptable practices, procedures, and principles that, when taken as a whole, have the effect of altering the ratio of undesirable events to total events, the first principle and the importance of the probability theory become self-evident. The problem that security must constantly deal with is that all undesirable events are breaches of security! The goal of security design is to decrease the ratio of unfavorable events to total events. Obviously, some events are more likely to occur than others in the same environment. The risk of a flood that inundates a city would seem less likely than a transient power failure (tell this to the people of New Orleans, LA). Both are undesirable events. Both can affect the operation of businesses. When the probability of each case is different, the ratios of favorable cases are added.

Two events that have no relation to each other are considered to be independent. If they are not linked in any way, the probability of their simultaneous occurrence is the product of their respective probabilities. An example: What is the probability of lightning striking a second time in the same spot? It is the same as the probability of its striking the first time: the two events are independent of each other. In security, the penetration of a system and the simultaneous failure of the security system from causes other than penetration may be expressed as the product of the probabilities of the independent events. This fits the condition of Laplace's principle 3, above. Many security (and safety) systems, such as those employed at nuclear power facilities, are based on redundancy, such that multiple failures must occur; the redundant systems do not become operational until the preceding systems have failed. Principle 4 expresses the relation between dependent events: the probability of the first event is multiplied by the probability of the second event if the second event can happen only after the first event has occurred. Breaking and entering followed by theft, to produce a burglary, is an example.

The probability of security system failure may be expressed in terms of the lower-risk multiple or backup systems. As for when two events are combined, principle 5 expresses the idea that when dealing with events, the past does not affect the future. If we assume the risk of a security breach is a given value and that it has occurred, we may not assume that it will not occur again. Probabilities of events are not guarantees. If an event has a probability of 1 in 100, the probability of that event happening again is still 1 in 100. For example, tossing a coin for heads or tails is a 50–50 proposition; that is, 1 time out of 2 it should come up heads. A coin toss could come up heads 10 times in succession; however, as the past cannot affect the present or the future, the chance of heads coming up each time is still 50–50, and it will remain so on every toss of the coin.

Principle 6 deals with the attribution of causes to effects. It describes the relation between all causes and probable causes. This is effectively the expression of circumstantial evidence, as a probability leading to a conclusion but one less convincing than the conclusion reached using direct evidence. Principle 7 involves the basis of confidence limits. To illustrate, if a random sample of 100 variables is taken and is found to have a mean of 40 and a standard deviation of 11, it will not be possible to determine a precise mean. The best that can be established is limits within which the mean will fall with a specified probability or confidence, usually taken as 95 percent. Again we need to ask ourselves the question, "How precise a measurement do we need?"

The definition of mathematical hope is essential to the design of a secure system. This concept relates the potential gain to the probability of obtaining the gain. Principle 8 allows the utility of a procedure to be expressed in both monetary and probabilistic terms. If the potential gain from a security system was $1,000 and the probability of achieving this gain was 1 in 500, a value of 2 (in arbitrary units) could be assigned. Equivalent values could be assigned to other combinations to allow comparison among alternatives. But why go to all this trouble for so little gain?

Principle 9 allows for the fact that any solution to a problem introduces risk. Risk management solutions may fail, and this must be considered in the design stage. A backup system to provide redundancy is certainly to be considered, as well as the cost/benefit ratio for doing so. This principle is extensively used in the manufacture of commercial aircraft and the space shuttle, for obvious reasons.

The condition to be considered last is the situation in which one of the alternatives to positive action is to do nothing. In some cases the risks, upon analysis, become insignificant; the decision may be to accept the possibility of loss because the potential loss will not have a substantial effect on assets. Principle 10 relates the amount and potential of risk to the wealth of the protected entity. A very profitable company may well afford to risk assets to maximize gains. The potential losses might be too great, however, for a less prosperous company, one that may be in greater need of relief from such occurrences. In some instances, then, the most cost-effective security is simply not to implement a plan or solution; in some others, the most practical solution might be to cover the potential loss with some form of insurance.

To summarize, risk can be expressed in terms of probability of occurrence. The goal of security system design is to improve the ratio of favorable events to total events, or to reduce the ratio of unfavorable events. The basic technique used is to rate risks on their probability of occurrence and to establish economic values for potential risks and potential solutions. When possible (and cost-effective), redundant or backup systems may be designed to provide the added dimension of protection. Risk probability is not a guarantee that, because an event has a low probability and has occurred only one time, it will not occur again.

Statistical analysis as it is used in many fields—astronomy, agriculture, engineering, or insurance—is approached by basically using the procedures enumerated above.

Again, a word of caution—no statistical procedure can, in itself, ensure there will be no mistakes, inaccuracies, faulty reasoning, or incorrect conclusions. The data must be accurate, the methods properly applied, and the results interpreted by someone with a thorough understanding of the field in which they are being applied. That, after all, is the hallmark of the professional.

ESTIMATING FREQUENCY OF OCCURRENCE

When experience (history) has provided an adequate database, loss expectancy can be projected with a satisfactory degree of confidence. For example, if one leaves the keys in the ignition of an unlocked car on a downtown street in a high-crime area, it is just a question of time until the car is stolen.

In new situations, however, or in situations in which data has not been or cannot be collected, we have insufficient knowledge on which to base our projections. An example would be the kidnapping of a high-risk-profile businessman in the absence of any prior threats or other indications that he had been targeted for kidnapping. In such instances, quantification of risk tends to be nothing more than educated guessing. It is in cases such as this that the services of an experienced security professional are needed to reduce subjectivity to an absolute minimum and to deal with the data available, limited though it may be, in a calm, objective manner. This is also true of international and domestic terrorism and, to a lesser degree, workplace violence. Amateurs tend to become very emotional—that is to say, less objective—when faced with such dangerous issues. The services of an outside consultant or a security professional experienced in such matters are essential in cases of this type, to ensure objective and proper analysis from the outset. Another example is in dealing with the threat of terrorism. Terrorists engage in seemingly random acts of violence. It is difficult, if not impossible, to estimate the frequency of occurrence of random acts of violence. By their very nature, they are highly unpredictable. So, we are left with the nagging thought that "risk analysis" is not an exact science, but an art, and we still have many challenges ahead of us, especially in dealing with terrorism.

4 ⬜⬜⬜
⬜⬜⬜
⬜⬜⬜

Quantifying and Prioritizing Loss Potential

"Security is more art than science. Few formulas will cover all organizations, situations and needs, and that's the beauty and challenge of our profession."
—Richard D. Sem, CPP, Former president, International Security Management Association

As with any complex chain of interrelated issues, overall strength is measured by the weakest link. Very strong security in one area will not compensate for very weak security in another. To proceed to correct conclusions and then to recommendations for corrective action, it is necessary to quantify and prioritize all the loss potentials identified. For the professional, here lies one of the most difficult tasks in the survey process—the task of measurement, or quantification, of exposures. Given adequate historical or empirical data, loss expectancy can be projected with a satisfactory degree of confidence. On the other hand, when there are insufficient data for reliable forecasting because the data either have not or cannot be collected, one is left with the nagging suspicion that conclusions will be nothing more than an exercise in educated guessing—not that there is anything wrong in that.

Many risks may be classified as things that might happen but that have not yet occurred. Such risks can either be accepted or minimized, using prescribed preventive measures. Acceptance assumes that the risk is not sufficiently serious to justify the cost of reduction, or that recovery measures will ensure survival, or that cessation of operations, if the risk should occur in its most serious magnitude, is an acceptable alternative. Minimizing the risk assumes that the risk exposure is or may be serious enough to justify the cost of eliminating or reducing the possibility of its occurrence, and that recovery measures alone will not always be effective in ensuring survival. Also, it postulates that the remaining alternative—cessation of operations—is unacceptable.

It is at this juncture that quantifying or prioritizing the loss potential becomes the hallmark of the true professional. We have often told clients that it does not take much talent to prescribe an 85 percent solution for a 15 percent problem. Real talent comes into play when one is able to diagnose client problems correctly and recommend necessary countermeasures to solve them without engaging in overkill. Granted, when we err, it must be on the side of prescribing more rather than less security, but not to the level that turns a college campus, hospital, or resort hotel into a *Stalag 17* prisoner-of-war camp.

There are always several tradeoffs when one considers the implementation of a new or improved security program. Cost is the most obvious. Less obvious and often overlooked are the inconveniences new security systems cause to personnel and the probable impact on employee morale. This is especially true if the employees perceive (rightly or wrongly) that the inconvenience caused them is greater than the threat. This type of "solution" can cause more harm than if management had done nothing at all. As an example, we observed that when excessive access-control systems were installed in a computer department of an airline reservations center, it resulted in employees' propping doors open for simplicity of movement during working hours. When queried, employees said, "The inconvenience was a bigger problem than unauthorized access."

ASSESSING CRITICALITY OR SEVERITY

Some authors refer to this stage of the survey as assessing criticality or severity of occurrence.[1] Regardless of what one calls the process, it is vital to search for and locate the proper benchmark to adequately approximate dollar values for the loss probabilities previously identified. Once this is done, the task of comparing the cure to the disease becomes self-evident. One can then develop a list of meaningful solutions with priorities based on a common denominator—the dollar. One technique in use is the three-stage approach, involving prevention, control, and recovery. Prevention attempts to stop undesirable incidents before they get started. Control seeks to keep these incidents from affecting assets, or, if impact occurs, to minimize the loss. Recovery restores the operation after assets have been adversely affected.

Many professionals take the approach that prevention is sufficient, and yet they opt for the installation of various control measures. It is one thing to install fire alarms that signal a serious situation; it is another to respond to and control a fire, and then recover from its effects. Similarly, it may behoove corporate management not only to have adequate security in place to prevent kidnapping attempts but also to design a contingency plan to deal with the kidnapping event, should preventive measures fail and the event become an actuality. In addition, nothing mentioned previously—prevention, detection, control, or contingency planning—precludes the necessity of having adequate insurance to help recover from a serious fire or successful kidnapping, extortion, and ransom event.

Another technique for assessing security is to prepare a segmented schedule of overhead, installation, and operating costs for the security project. All costs identified must be directly chargeable to expected benefits. In this process, it is crucial to show that the benefits (risk prevention or reduction) will outweigh the cost. This is referred to as a *cost/benefit summary* and is useful for both existing and proposed security programs and projects. (This will be more fully discussed in Chapter 5, Cost/Benefit Analysis.)

---

[1]T. J. Walsh and R. J. Healy, *The Protection of Assets Manual* (Santa Monica, CA: Merrit, 1974).

Table 4.1 Decision Matrix: A Risk-Handling Decision Aid

	Frequency of Loss		
Severity of Loss	High	Medium	Low
High	Avoidance	Loss prevention and avoidance	Transfer via insurance
Medium	Avoidance and loss prevention	Loss prevention and transfer via insurance	Assumption and pooling
Low	Loss prevention	Loss prevention and assumption	Assumption

THE DECISION MATRIX

Another simple technique for prioritizing loss potential is the use of a frequency and severity loss matrix as an aid in making decisions about handling risk. Table 4.1 uses the adjectives *high, medium,* and *low* as factors to measure both frequency and severity of loss.

The quantification and prioritizing of loss potential should take into account the fact that there are both "intuitive" security control concepts, such as the installation of a burglar alarm at a warehouse, and security control concepts based on detailed cost/benefit analysis. An example of the latter is a multiple-stage electronic card-access control system for the research and development laboratory of a computer chip manufacturer. The procedures for both approaches take into full consideration the following:

- Available information resources
- Reliable probability relationships
- Minimum time and resource requirements and availability
- Maximum incentives for management cooperation
- A realistic evaluation of existing or planned security control effectiveness

The means of protection designed must always be tailored to the specific risk in the real day-to-day working environment of the specific entity being studied. The application of controls simply because they are recommended by some vague standard or acceptable practice, without regard to risk in the real-world environment, often results in controls that are inappropriate, ineffective, and costly. Worse, as so often seen with inappropriately planned closed-circuit television (CCTV), such controls may generate a false sense of security on the part of management.

For example, I was once asked to review the installation of a CCTV security system for a newly constructed newspaper plant in California. The CCTV system had been designed for the corporation by a building and facilities engineer. The plant was located in a newly developed industrial park. The CCTV system had apparently been planned without regard to environmental considerations. Upon review, it was determined that the CCTV system—complete with zoom, tilt, and pan lenses as well as video cassette tape-recording functions, all very costly to install and maintain—was operating in an area that had heavy fog about 6 months of the year during nighttime hours. Further, the fence line, at its nearest point to the building, was 350 yards away from the closest camera lens! To the question, "Why install CCTV at this location?" the answer was,

"We have used CCTV successfully at all our other plants and it just seemed the natural thing to do here."

They did have CCTV at their other plants, but in the plants that I inspected, the CCTV systems were more often than not inoperative, in whole or in part, because of inadequate maintenance and repairs needed on cameras, monitors, and video recording units. The CCTV systems were regarded by operations personnel as expensive toys that added little to the security of the facility. Management, however, was proud of its security program, having been lulled into a false sense of security by the presence of the CCTV cameras.

The assessment of risk, using actuarial methods to handle large numbers of events or situations, will be further examined in the following chapter. This technique has become generally reliable. However, in our experience, the entire exercise of estimating risks for a specific installation or complex is at best imprecise. Defining risks by using highly specific numbers has not always been validated by experience. Several well-known authorities have concluded that order-of-magnitude expressions, such as low, moderate, and high, to indicate relative degrees of risk are more than adequate for most risk-control surveys.

The terms *low, moderate,* and *high* equate roughly with probability ranges of 1 to 3, 4 to 6, and 7 to 10, respectively. One is cautioned here to remember that even a low risk should be taken seriously if the potential damage (or danger) is assessed as being moderate to high. An example would be the kidnapping for ransom of a high-profile business executive in a foreign country. We may regard the risk to be nearly nonexistent (low), but the potential danger and the impact on the company of such an unfavorable event should always be rated as high.

5

Cost/Benefit Analysis

"Violence assessment, like all forms of risk assessment, guides the use of limited resources (time, budget and personnel) to maximize benefit."
—James S. Cawood, CFE, CPP, CPI, *Violence Assessment and Intervention: The Practitioner's Handbook,* CRC Press, 2003

When we use the systems approach to conduct security surveys, problems are properly identified, analyzed, and quantified in terms of the seriousness of their impact on the operation or facility. Only solutions specifically responsive to a demonstrated need or requirement are considered. Only those tools or techniques that perform the needed task most effectively at the least possible cost are designed and then introduced into the system. Efficiency versus cost is, then, the first phase of balancing the cost/benefit ratio, which is essential to the proper development and design of effective security counter-measures.

The following techniques are increasingly being used to analyze, develop, and design cost-effective security. This includes policy, procedures, hardware (electronics), and manpower utilization programs.

SYSTEM DESIGN ENGINEERING

If a facility has a security program, a good systems engineer can review it and make recommendations to consolidate, coordinate, upgrade, and improve the existing protection. If a security program does not exist, the systems engineer can design one that properly marries the best of the following into a comprehensive and cost-effective security operation:

- Written policy, procedures, and guidelines
- Hardware (lights, lock and key controls, card access, anti-intrusion alarms, to name a few)
- Manpower (guard service or proprietary security personnel)

This review can be accomplished by using any one of the many comprehensive review techniques available, such as the one set forth in Appendix A, Security Survey Work Sheets. By asking questions and directly observing certain operations, such as the adequacy of lights at night in an employee parking lot adjacent to a plant, one can reach certain definite conclusions. Elements of the security program (lights, locks, alarms, guard coverage, and so forth) are adequate, inadequate, or nonexistent. Obviously the two former situations are a subjective matter, whereas the latter leaves no room for argument.

Other, more sophisticated programs use advanced electronic techniques. An example is electronic filtering as applied to access control in security programs for highly sensitive environments, such as research and development facilities. (For an example of an electronics security system specification, see Appendix H.) Additionally, we are seeing ever-increasing use of computer programs with models to conduct cost/benefit ratio and computer-aided design (CAD) analysis. Whatever technique is used to determine the cost/benefit ratio, three basic criteria should be considered before the proper procedures or countermeasures are selected: cost, reliability, and delay.

Cost

Initially, one tends to focus primarily on the acquisition cost. We must, however, also take into consideration life cycle and replacement cost factors as well. For example, the initial base cost to re-core and re-key the locks of an entire facility may be $15,000, plus 10 percent per year for inflation, or $22,500 after a lapse of 5 years.

For a key-and-lock system that includes changeable cores and an in-house capability for cutting keys, the initial capital outlay of $15,000 may well be amortized over 15 to 20 years instead of the original 5 years. Here we have considered all three factors: acquisition cost, the life of the system, and replacement cost. The cost of contract versus proprietary security guard service, or a combination of the two, can also be calculated, using the same principles. The same exercise can be accomplished for any type of hardware or electronics equipment with only a slight variation in computation.

Reliability

Reliability is especially critical with hardware and electronic devices, such as anti-intrusion and computerized card-access control systems. The state of the art in electronics systems is advancing faster than most people can imagine, much less keep track of. Consequently, components are being manufactured, sold, and installed before being properly field-tested. This inevitably leads to difficulties—difficulties that, unless corrected immediately, translate into expensive electronics systems designed to solve problems that, in turn, create new problems, which are often difficult and time-consuming to correct.

I know of only one solution to this problem from dragging on for months: build two written requirements into the contract of proposal or purchase. First, require the successful bidder to present for your inspection a site demonstration, a like system or unit installed on a property with the same or similar security or access-control needs. It should be one that has been functioning for a minimum of 6 months without problems. Second, include a clause in the purchase contract withholding the last payment (payments are usually made upon signature [one third], upon installation [one third], and at final system acceptance) until the system has been on line for a sufficient test period, say 90 days, to ensure that the product is problem free and all bugs have been located and eliminated. Such a clause will make it in the supplier's best interest to get the unit fully operational as quickly as possible.

A word of caution: no legitimate hardware or electronic security systems supplier will object to the inclusion of the above protective clauses. A supplier that balks at either or both cannot or will not guarantee satisfactory results for his or her system. This being the case, it would probably be in the best interest of the buyer to look for a supplier who will. Nothing is sadder to behold than a well-meaning director of security who has finally convinced management of the necessity for an expensive security system that, after acquisition, continually malfunctions. Although cost, like the state of the art, is ever changing, these sophisticated, computerized systems can be extremely expensive.

Delay

Time is the third factor to consider. How long will it take, in comparison to other countermeasures that could be used, before the recommended system can become fully operational? Here we may well have to consider the possibility of having more than one countermeasure operating at the same time until the primary system renders the secondary system obsolete or less critical.

An example is the introduction of a multilevel, electronic anti-intrusion and card-access system in a facility that houses separate functions under one roof. Some of these functions—such as the research and development laboratories, and the storage area in a warehouse, which contains a $6 million inventory of easy-to-conceal and easy-to-sell items—may require differing levels of protection at different times.

It is probably going to be necessary to incur manpower costs for a guard force to secure the premises 24 hours a day, 7 days a week, until the new access system becomes fully operational. This amounts to 168 hours of guard service coverage. If each guard works a 40-hour shift, it will take 4.2 guards (taking into account time for relief, sick time, vacation, and so forth) to accomplish this task. If the cost for this service is $9.50 per hour, the total cost is as follows:

$$168 \text{ hours} \times \$9.50 \text{ per hour} = \$1,596.00 \text{ per week}$$
$$(\$228.00 \text{ per day per post}),$$

or

$$\$6,840 \text{ per month, or } \$83,220 \text{ per year (365 days)},$$

or

$$\text{average monthly cost } \$7,068[1]$$

---

[1]The average monthly cost is arrived at by dividing annual dollars by 12 months. For more specific and detailed costing, the daily rate times the number of days in a given month may be used. Thus, $228 \times 31 = \$7,068$. For accounting purposes, one can use the same formula to compute the cost of any guard service contract. According to a Bureau of Labor Statistics, U.S. Department of Labor study (Bulletin 25 40-11, 2002), contract security guards earn between $7.80 and $11.43 per hour.

Needless to say, a prudent, cost-conscious director of security will make every effort to phase-in the anti-intrusion card-access system and phase-out or reduce the manpower requirements as quickly as possible.

BUILDING REDUNDANCY INTO THE SYSTEM

To achieve very high levels of reliability in security programs, one should consider building redundancy into them. An example is the use of multiple smoke sensors in a hotel or housing areas to warn of the incipient stages of a potentially disastrous fire. The old saying, "One may well survive a burglary, but a good fire can put you out of business forever," has much meaning for a hotel complex or a computer library. If we install multiple-use smoke detector sensors (ones that use all three detection techniques—ionization, infrared, and the photoelectric cell), the chance of all their modes failing at the same time is statistically 10,000 to 1. Yet in terms of cost, these units can be generally obtained and installed for about $250 each—a small outlay compared with the cost of replacing even the least expensive computer equipment, not to mention the potential catastrophe of the loss of life or of losing the materials stored in a computer library as a result of a fire that could have been easily detected.

Redundancy can also be accomplished by designing a proprietary alarm system, which for a few dollars more can be remotely monitored by a central alarm station as an added backup, against the possibility of a power or human failure at the facility's proprietary alarm console. The costs—a dedicated lease line and a rental (monitoring) fee—are relatively inexpensive.

In art museums, multiple redundant systems usage is a common technique. The items on display in an art museum must be readily available for the viewing public; thus, security must be as unobtrusive as possible. Many of these art objects, however, are one-of-a-kind treasures and are therefore priceless. Security in the daytime, functional though minimal, may include uniformed security guards, closed-circuit television, antipenetration display case alarms (local, audible, and remote), and antitamper or antiremoval switches behind picture frames or on wall mounts. For nighttime security, additional multiple anti-intrusion sensors and motion detectors may be employed to ensure that if an unauthorized penetration of the perimeter barriers happens to occur (such as a remain-behind burglar), the movement and the presence of such a person inside the museum will be detected by multiple sensing devices. Any one of them can be circumvented or might not be fully operational; nevertheless, the odds against all of the devices failing at the same time and therefore not detecting the presence of an intruder are statistically in the thousands, and therefore the total failure of all systems is virtually impossible. It should be noted here that many art museums have systems installed that are designed to capture an intruder inside the museum, thus negating the possibility that a burglar would accomplish the theft and then successfully flee the premises before the security force can react.

The selection of the right countermeasure to control or minimize the identified risks will take into consideration the fact that written policy, procedures, and guidelines are less expensive than hardware but that hardware (including electronics) is less expensive than manpower. So when looking for the proper "fix," it is well to start with the basics and then work one's way up to the more sophisticated and thus more costly fixes.

As an example, there may be no adequate substitute for the use of mechanical equipment, fences, gates, locks, safes, and vaults. In some locations, however, local real estate codes, covenants, and restrictions (CC&Rs) may prohibit the installation of a 7-foot chain-link fence with the usual three-strand barbed-wire top overhang. In these instances, not uncommon in industrial parks, one must retreat to the exterior wall of the building or buildings as the place to begin perimeter security. It may then become necessary to use electronics, alarms, a computerized console, closed-circuit television, and an exterior and interior security guard patrol, among other things, to provide the needed security.

A SECURITY COUNTERMEASURE

If the previously described problem is encountered in a new industrial park, one possible solution is to prepare the security countermeasure plan in stages or increments, such as stages 1, 2, and 3, or increments A, B, and C. In this technique, we "cost out" the use of each of the required systems in terms of the minimum level of security that one or more countermeasures will provide. Then, we move up to stages 2 and 3, adding more security countermeasures to the building or complex, adding ever-increasing cost to the project's countermeasures plan.

This type of program is relatively easy to explain and hence to sell to management. The underlying philosophy is, "We will try stage 1 or increment A first, at $X. Should stage 1 (increment A) fail to provide the needed level of security, we will move to stage 2 (increment B), and so on, until we solve the problem." This technique will prevent the all too commonly encountered security "overkill" situation. A security risk properly accessed in the 15 percent range does not need an 85 percent solution, nor can management afford one. Yet given a brand-new environment, such as a recently developed industrial park in a suburban area, who can say exactly what level of security will be needed? In the absence of empirical knowledge to the contrary, our systems countermeasure design technique is both efficient and cost-effective. What it says to top management is, "We can start out with the basics, which are the most effective for the least money, and then add to the system (at greater cost) as we develop the necessary historical data to justify spending more money."

Most cost-effective security systems use a combination of procedures, hardware (electronics), and manpower to achieve the proper countermeasures balance. The first edition of *Risk Analysis and the Security Survey,* in 1980, made the following prediction:

In the next decade we will see more use of security systems that integrate many separate functions, systems that are developed, manufactured, sold, installed, and maintained by one company. These systems will integrate security, communications, fire, life safety, building management, and energy control from one central console or command control center. The big users will be high-rise office buildings, retail stores, and shopping malls. Specialized units will be developed for use at airports, oil refineries, and electronics manufacturing plants, to name a few.

This prediction became an actuality within 5 years. With increased attention now being paid to security as a result of the events of 9/11, it is anyone's guess where we will be 10 years from now.

In conclusion, even the best-designed countermeasures system must be proved cost-effective before it can be sold to management. Only by reducing or integrating the largest cost factor, manpower, and replacing it where practical with procedures, hardware, and electronics can we achieve more effective security at less cost. This is a proven technique. Finally, whenever dealing with security vendors, whether for manpower, hardware, or electronics, one should obtain a minimum of three bids based on a written specification of requirements. Not only are competitive bids a proven cost-effective technique, but they also tend to keep everyone honest. Bidders should be advised of the competition, but not necessarily the identities of the other competitors. Sole-source procurement is seldom cost-effective and more often than not provides fertile ground for financial manipulation, which is seldom in the client's best interest. In conclusion, one should always keep in mind one of Charlie Hayden's cardinal rules for conducting security surveys: "The optimum reduction of Risk occurs at that point at which further reduction would cost more than the benefits to be gained."

6

▢▢▢
▢▢▢
▢▢▢

Other Risk Analysis Methodologies

Eugene Tucker, CPP, CFE, CBCP
Contributing author

"He kissed me and our young daughter good-bye. I never predicted or imagined that he was involved in such horrific activities. He had a kind, caring and calming presence about him."
—Samantha Lewthwaite describing her husband Germaine Lindsay, a suspected suicide bomber who killed 26 people in the London King's Cross attack.

Before September 11, 2001, President Clinton signed Presidential Directive 63, the Policy on Critical Infrastructure Protection. It identified 8 (now 11) sectors of the economy considered critical to national security. Included are telecommunications, transportation, water supply, oil and gas production, banking and finance, electrical generation, emergency services, and essential government functions. This Directive, along with the Bio-terrorism Act and other implementing policies, assigned oversight of each function to a separate governmental agency. The protection of the water supply is the responsibility of the Environmental Protection Agency (EPA); the protection of the food supply is the responsibility of the Food and Drug Administration (FDA). These agencies are assigned the task of developing risk assessment and security protocols for the protection of the assets under their purview, with many using a different risk assessment methodology.

Many risk and vulnerability analysis methods exist. Although similar in nature, security professionals should be aware of the basics of these differing methodologies even if they are not involved directly in the function they assess. In this chapter, I briefly outline methodologies resulting from the above: Operational Risk Management (ORM), CARVER + Shock, and VSAT a spin-off of Risk Assessment Methodology—Water (RAM-W).

VSAT

VSAT is an acronym for the Vulnerability Self Assessment Tool and is both a methodology and software tool used to develop security systems capable of protecting specific targets from the acts of specific adversaries. As such, it can be considered a qualitatively based (asset-based) methodology. Its stated goals are to assess vulnerabilities, develop priorities based on the cost and feasibility of remediation, and determine potential solutions for the prioritized vulnerabilities. Although developed for water and wastewater systems, it can be used for assessing the vulnerability of other process-intensive systems. The software produces standardized reports and organizes the vulnerabilities into a color-coded threat matrix. The software also contains a library of typical water system assets, security threats, and countermeasures to help nonsecurity professionals complete the analysis. It allows the user to modify and define additional threats and countermeasures.

VSAT uses a baseline assessment and security improvement analyses to compute Risk Reduction Units for the countermeasures assigned by the analysis. The cost of the countermeasures is also calculated, resulting in the cost per reduced risk and allowing management to support decisions related to security improvements.

There are 11 steps in the VSAT assessment process:

1. Identify assets
2. Identify threats
3. Determine criticality
4. Identify existing countermeasures
5. Determine risk level
6. Determine the probability of failure
7. Assign vulnerability
8. Determine whether risk is acceptable
9. Develop new countermeasures
10. Perform risk–cost analysis
11. Develop a business continuity plan

Identify Assets

The first step requires the analyst to identify and list all assets into five categories:

- Physical
- People
- Knowledge base
- Information technology
- Customers

Identify Threats

An assessment is made about whether a range of manmade and natural[1] threats pose a risk to each individual asset. The extent of the threat posed to the asset is also considered at this juncture. This should give the analyst a general sense of the system's vulnerabilities.

---

[1]The identification and assignment of technological threats are not mentioned in the methodology but should almost always be considered for a broader and more complete vulnerability perspective.

Determine Criticality

Each of the asset characterizations identified above is assigned a criticality level of low, moderate, high, or very high.

Identify Existing Countermeasures

At this point, existing countermeasures (physical security and other controls) are identified and evaluated to help understand the level of vulnerability.

Determine Risk Level

Each vulnerability is evaluated using a two-dimensional matrix, but the number of rating levels can and should be increased as necessary.

Determine the Probability of Failure

What is the likelihood of the failure of existing countermeasures when pitted against the variety of threats? This category looks at the effectiveness of controls in place, including the ability to mitigate, respond, and recover from threats and asset (critical system component) failure.

Assign Vulnerability

Vulnerability is defined as the probability of failure and the probability of occurrence after countermeasures are implemented. It measures the likelihood of the threat and its ability to cause damage. It is intended to be qualitative and subjective by assigning general vulnerability levels that include very high, high, medium, and low. The ranges must be defined in the context of the specific environment.

Determine Whether Risk Is Acceptable

Levels of risk acceptability are defined at this phase and assigned a color code. Red can represent unacceptable risks; green represents a willingness to accept the risk. Yellow is between the two and defines risks that are unacceptable but that management chooses to accept.

Develop New Countermeasures

As a result of the threat vulnerability analysis, determination is made of what new countermeasures are required to reduce the risks.

Perform Risk–Cost Analysis

Several cost/benefit methods are used to determine which countermeasures will return the greatest value. The program uses a simple cost method that returns the annualized costs through the sum of the annual operating costs plus the capital costs divided by the useful life of the countermeasure. This method does not consider the time value of funds or countermeasure funding and installation.

The Debt Payment method considers the interest paid on capital investments and returns an annualized cost.

The VSAT software prioritizes countermeasures based on the anticipated risk reduction per dollar spent.

Develop a Business Continuity Plan

The final step is to develop plans for improvements. A Business Continuity Plan per se is not produced; it is a plan for improvement activities needed to mitigate or manage the known risks.

The process is intended to return a series of reports that allow management to trade off cost against risk, enabling them to mitigate risks at the lowest cost.

OPERATIONAL RISK MANAGEMENT

The FDA is responsible for the security of food production, importation, warehousing, transportation, and distribution in the United States. This agency has adopted the Operational Risk Management (ORM) method to help ensure that food assets remain safe from attack.

ORM is an engineering-based risk management system used by the Federal Aviation Administration and the military to examine the safety and risk to existing systems. It is a tool designed to help identify operational risks and benefits to determine the best course of action for any given situation. It is defined as the risk of loss resulting from inadequate or failed processes, people, and systems or from external events. It is similar to other risk management methods in that it identifies hazards and determines the impact of food safety through estimating the probability and severity of an attack, allowing those responsible to focus on the worst hazard first. The FDA defines it as a defensive vulnerability assessment tool to identify points in a system that are most susceptible to terrorist attack and to design preventative measures to reduce risks. It uses a six-step process:

1. Identify the hazards
2. Assess the risk
3. Analyze risk control measures
4. Make control decisions
5. Implement risk controls
6. Supervise and review

Identify the Hazards

The first step examines each activity in a process or flow of actions and events and identifies the associated hazards. ORM defines a hazard as any real or potential condition that can cause degradation, injury, illness, loss of equipment, or property damage. In the case of restaurant food security, a process may include the placement of food in a salad bar. The associated hazards include intentional contamination with bacteria, the placement of sharp objects inside the food, or the introduction of a poison.

Assess the Risk

Hazard probability, severity, and exposure (the number of people or resources affected by a given event or cumulative events) are determined in terms of their impact on people and food security. ORM uses the probability categories of frequently, likely, occasional, seldom, and unlikely. In the context of food security, probability is defined as follows:

- Frequently: takes place often, and persons are continuously exposed
- Likely: takes place several times, and persons are regularly exposed
- Occasional: will happen, and exposure is sporadic
- Seldom: may happen, and exposure is infrequent
- Unlikely: likelihood and exposure are rare

Severity is categorized as catastrophic, critical, moderate, or negligible:

- Catastrophic: complete business failure or loss of facility asset due to attack resulting in fatalities
- Critical: major business impact resulting from severe illnesses or incident
- Moderate: minor business impact resulting from minor illnesses or incident
- Negligible: less than minor business impact or illness

Risks are then ranked using a matrix.

Analyze Risk Control Measures

This step represents the development of risk control measures to mitigate, prevent, control, or eliminate the hazard or reduce its probability or exposure.

Make Control Decisions

In the ORM methodology, decision makers who can implement the control measures are identified. ORM does not assume this is a singular person or department.

Implement Risk Controls

Once the controls and those responsible for the implementation of the controls are identified, the process of executing protective measures begins. Implementation strategies are developed that define individual responsibility, accountability, and involvement. ORM allows for different people to implement the controls based on severity.

Supervise and Review

Once controls are in place, their effectiveness is examined and reevaluated. Additional assessment is completed as the system changes.

The following principles apply to all stages of the ORM process:

1. Accept no unnecessary risk. Unnecessary risk comes without a commensurate return in terms of benefits or opportunities.
2. Make risk decisions at the appropriate level. Making risk decisions at the appropriate level establishes clear accountability. The decision maker is the person who can allocate the resources necessary to implement controls or is authorized to accept risk.
3. Accept risk when the benefits outweigh the costs. All identified benefits should be compared with all identified costs. Balancing costs and benefits can be a subjective process.
4. Integrate ORM into planning at all levels. The later changes are made in a process of planning and executing an operation, the more expensive and time-consuming they can become.

CARVER + SHOCK

The U.S. Department of Agriculture (USDA) Food Safety and Inspection Service (FSIS), in conjunction with the Homeland Security Office of Food Security and Emergency Preparedness (OFSEP), has adapted the U.S. Department of Defense's military targeting method called CARVER + Shock to reduce or eliminate the potential risk at vulnerable points along the "farm-to-table" continuum. CARVER identifies vulnerabilities from a terrorist's point of view.

The USDA defines CARVER + Shock as an offensive target prioritization tool to identify critical nodes most likely to be targets of terrorist attacks and to design preventative measures to reduce risk. It is used in conjunction with ORM.

CARVER + Shock considers seven factors that affect the attractiveness of a target:

- Criticality: the degree to which the public health and economic impacts achieve the attacker's intent. How important is the target as determined by the impact of its destruction on operations?
- Accessibility: physical access to the target. How easily can a target be reached, either by infiltration or with weapons?
- Recuperability: ability of the system to recover from an attack. How long will it take to replace or repair the target once it is damaged or destroyed?
- Vulnerability: ease of accomplishing an attack
- Effect: amount of direct loss from an attack
- Recognizability: ease of identifying a target
- Shock: psychological effects of an attack

Management then develops their security strategy based on the results of the analysis.

7

The Security Survey: An Overview

"A security survey is a critical, on-site examination ... to ascertain the present security status, identify deficiencies or excesses, determine the protection needed, and make recommendations to improve the overall security of the operation."
—Raymond M. Momboisse, *Industrial Security for Strikes, Riots and Disasters,* Charles C. Thomas Publishers, 1977.

The goal of risk management—to manage risk effectively at the least possible cost—cannot be achieved without eliminating or reducing, through a total management commitment, the number of incidents that lead to losses.[1] Before any risk can be eliminated or reduced, it must first be identified. One proven method of accomplishing this task is the security survey. Charles A. Sennewald, author and security consultant, has defined the security survey as follows: "The primary vehicle used in a security assessment is the survey. The survey is the process whereby one gathers data that reflects the who, what, how, where, when and why of the client's existing operation. The survey is the fact-finding process."[2]

WHY ARE SECURITY SURVEYS NEEDED?

The latest reports estimate that the cost of fraud and abuse to American business is in excess of $600 billion per year and rising. This figure is believed by most authorities to be very conservative. The biggest problem, and the one we see most often, is that most corporate managers do not know if they have problems. Worse, many do not even want to know that they have a problem! Some managers seem to prefer to keep things as they are and to regard any suggestion of the need for increased security as direct or indirect criticism of their ability to manage their operation. I hope that this attitude has changed

---

[1] The field of risk management encompasses much more than security and safety. These two subjects, along with insurance, however, are the cornerstones of most effective risk management programs.

[2] C. A. Sennewald, CPP, *Security Consulting*, 3rd ed. (Boston, MA: Butterworth-Heinemann, Elsevier, 2004).

for the better since 9/11. Only time will tell. Nevertheless, where fraud exists, most business fraud surveys calculate losses at 6 percent of annual revenue. Some surveys we have conducted have concluded that losses attributable to theft equaled or exceeded profits! This is especially true for chain-store operations, in which each individual store is regarded as a separate profit center, and records of inventory shortages are kept for each location by local managers.

Crime losses far exceed the losses to business caused by fire and industrial accidents. One professional security organization estimates that the annual loss to business from fraud and abuse is twice as great as the total of all business losses due to fires and accidents!

Are you really concerned about how crime may affect your business? Take a few minutes and read the Association of Certified Fraud Examiners *2002 Report to the Nation on Occupational Fraud and Abuse.* Then take a moment to reflect on the following estimates:

- The loss of 6 percent of revenue translates to a loss of approximately $600 billion annually or about $4,500 per employee per year.
- Occupational fraud most often falls into one of three categories: asset misappropriation, corruption, or fraudulent statements.

In the United States alone, estimates of the cost of fraud vary widely. No recent comprehensive studies could be found that empirically measured the economic effects of fraud and abuse. In addition to the direct economic losses to an organization from fraud and abusive behavior, there are indirect costs to be considered: the loss of productivity, legal action, increased unemployment, government intervention, and many other hidden costs.

WHO NEEDS SECURITY SURVEYS?

The same study cited above shows that fraud is distributed fairly evenly across four broad sectors: government agencies, publicly traded companies, privately held companies, and nonprofit organizations. In our own experience, many victims were seen to be growing businesses in which expansion had occurred faster than the development of internal control systems to prevent fraud. These were relatively large companies, in which close control over divisions and branches were seldom established. This fact has been noted by many professional security consultants with whom I have discussed the subject. Every business entity, no matter how large or small, could profit from an objective survey of their security protection. Many surveys we have conducted showed that the majority of business security concern is directed toward external problems, such as theft (burglary and robbery), as the company's most immediate priority. These priorities (along with fire protection) are seen as primarily insurance driven. This situation reflects the development and growth of the security industry in the United States from a historical perspective: it was once thought that most, if not all, of a company's problems with theft were external in nature.

Up until World War II, before the large-scale expansion of U.S. and Canadian industry and the development of the multinational and international corporations, many companies in North America were what we now regard as small to medium-sized industries and businesses. Most retail enterprises were of the "Mom and Pop" variety. Because of close supervision and the personal identification between management and labor that existed, internal theft (theft by "trusted" employees) was seldom encountered. Hence, the fire and burglar alarm business was developed and rapidly expanded to protect these enterprises from what were then considered the only two threats to the profitability of the business—fire and burglary—their two "worst enemies."

With the growth of the national, multinational, and international corporations and the almost total demise of the "Mom and Pop" commercial and variety stores, we witnessed a parallel change in business ethics and standards. These changes affected society as a whole. As an example, in the turbulent 1960s, a new term was coined: *the Establishment.* Crimes against the Establishment were then, and still are, perceived by many to be permissible, in fact not crimes at all. The lack of personal identification with a company by its employees and the dramatic dilution of ethical and moral standards (Enron and WorldCom, for example) among management, employees, and the general public combined to make internal theft by employees a simple process of rationalization. After all, who is Enron, Toyota, or Safeway? "They" make gigantic profits by "ripping off" the general population (read "us"). The theft by an employee of one small item (or dollar amount), multiplied 1,000 times each year in as many companies, can add up to an annual loss that can far exceed the total loss attributed to external theft by a company over the entire period of a firm's existence!

Enlightened students of criminology have come to understand that the most predominant and certainly the most prevalent and most costly form of business crime is employee or internal theft in its various forms. Asset misappropriation accounts for the majority of employee theft. Bribery (corruption) and fraudulent statements, although not as frequent, run second and third. Companies with 100 or fewer employees are the most vulnerable to fraud and abuse.[3] Shoplifting, employee theft, and vandalism all cost American businesses billions of dollars annually. Table 7.1, based on percentage of a company's net profit, illustrates how much more a company needs to sell to offset losses in stolen merchandise, equipment, and supplies. The theft of one $500 item means the company must sell $25,000 worth of merchandise to break even, if it is in the 2 percent net-profit category, $8,333 if it is in the 6 percent category. Worse yet, it is still without the stolen item, which needs to be replaced! For an example of the impact of loss versus net profits, see Table 7.1.

ATTITUDE OF BUSINESS TOWARD SECURITY

In general, we find that most businesses will take the necessary precautions to protect themselves against the entry of burglars and robbers onto their premises. Most will also give protection to high-value areas, such as computer centers, vaults, precious-metal storage areas, and any location where money is the principal product, such as in banks and

---

[3]Association of Certified Fraud Examiners, *Report to the Nation*, 2002. A million dollars of revenue per year are particularly hard hit by crime. This is especially true for retail enterprises that operate on a 2 percent profit margin. As serious students of this problem have come to realize, losses due to crime can and do have a dramatic impact on net profits.

Table 7.1   Loss to Sales (Profit) Ratio

	If Company Operates at Net Profit of:				
	2%	3%	4%	5%	6%
Actual Loss of:	These additional sales are required to offset an actual loss:				
$ 50	$ 2,500	$ 1,666	$ 1,250	$ 1,000	$ 833
100	5,000	3,333	2,500	2,000	1,666
200	10,000	6,666	5,000	4,000	3,333
250	12,500	8,333	6,250	5,000	4,166
300	15,000	10,000	7,500	6,000	5,000
350	17,500	11,666	8,750	7,000	5,833
400	20,000	13,333	10,000	8,000	6,666
450	22,500	15,000	11,250	9,000	7,500
500	25,000	16,666	12,500	10,000	8,333

gambling casinos. Many businesses, however, still do not concern themselves with protection against unauthorized access to their premises—yet 75 to 85 percent of external theft is directly attributable to this source. A prime example is unrestricted access at many warehouses we have surveyed to the shipping and receiving docks by nonemployee truck drivers.

Likewise, if the cost of security protection is regarded as a capital expense or a yearly expenditure, and as reducing the profit line, we can expect a person not educated in risk management in charge of a facility to authorize only the minimum and least expensive security protection necessary. If, on the other hand, security is mandated by senior corporate management or required by clients or the various governmental agencies overseeing industries engaged in sensitive government contracts, we find plant managers installing security protection, often with little regard to costs (nuclear power generating plants, for example).

This latter proposition presupposes another problem. Most plant managers with whom we have worked do not have the foggiest idea what kind of security they need for adequate protection. For sensitive government contract work, security guidelines are often mandated. Nevertheless, many are the managers who have fallen prey to the sales pitches of security manpower and hardware salesmen who proposed package deals that are guaranteed to solve all problems. As a result, we constantly encounter security programs that overemphasize manpower (guards) or hardware (alarms, closed-circuit television [CCTV], or electronics) when the proper solution to the problem is an effective marriage of the hardware and electronics with adequate written procedures to deal with the most common occurrences or eventualities encountered in the real, work-a-day world. Life becomes complicated only if we allow it to. Most security solutions are relatively simple in concept and application.

WHAT CAN A SECURITY SURVEY ACCOMPLISH?

One approach to determining whether there is a need for a security survey is to find out what services an experienced security expert can provide and then to seek information on security-related (crime-related) losses being incurred by the particular business or company. The security expert can be in place (that is, a member of the staff), an outside consultant, or a combination of both.

Security-related problems might reflect any of the problems mentioned in the early chapters of this book. Surveys generally, however, show that most problems encountered in the real world of security are not uncommon ones.

If the company or facility being considered has a security plan, the survey can establish whether the plan is up-to-date and adequate in every respect. Experience has shown that many security plans were established as the needs of the moment dictated; most were developed without regard to centralization and coordination. Upon review, many such plans are patched-up, crazy-quilt affairs. More often than not, the policies, procedures, and safeguards need to be brought together for review and consolidated so that the component parts of the plan complement, not contradict, one another.

If the facility has no security plan, the survey can establish the need for one and develop proposals for some or all of the security services commonly found in industrial settings. By conducting a comprehensive survey of the entire facility—its policies, procedures, and operations—one can identify critical factors affecting the security of the premises or operation. The next step is to analyze the vulnerabilities and recommend cost-effective protection. The survey should also recommend, as the first order of business, the establishment of sound policies and procedures, as a minimum to define the security mission and to accomplish the following:

- Protect against internal and external theft, including embezzlement, fraud, burglary, robbery, industrial espionage, and the theft of trade secrets and proprietary information.
- Develop access-control procedures to protect the facility perimeter as well as computer facilities in other sensitive areas, such as executive offices.
- Establish lock-and-key control procedures.
- Design, supervise, and review installation of anti-intrusion and detection systems.
- Establish a workplace violence program to help corporate personnel deal with internal and external threats.
- Provide control over the movement and identification of employees, customers, and visitors on company property.
- Review the selection, training, and deployment of security personnel, proprietary or contract.
- Assist in the establishment of emergency and disaster recovery plans and guidelines.
- Identify the internal resources available and needed for the establishment of an effective security program.
- Develop and present instructional seminars for management and operations personnel in all the above areas.

This list is by no means all-inclusive. It does, however, set forth some of the most basic security programs and systems recommended for development by security surveys that I have conducted.

WHY THE NEED FOR A SECURITY PROFESSIONAL?

Losses due to all causes continue to represent a problem of major proportions for business and industry. To the extent that the services of security professionals can help in eliminating, preventing, or reducing a company's losses, they are needed. At a security management seminar, Charles A. Sennewald once explained, "Crime prevention, the very essence of a security professional's existence, is another spoke in the wheel of total loss control. It is the orderly and predictive identification, abatement, and response to criminal opportunity. It is a managed process which fosters the elimination of the emotional crisis response to criminal losses and promotes the timely identification of exposures to criminality before these exposures mature to a confrontation process." The proper application of protection techniques to minimize loss opportunity promises the capability not only to improve the net profits of business but also to eliminate or reduce to acceptable levels the frequency of most disruptive acts, the consequences of which often exceed the fruits of the crime.

HOW DO YOU SELL SECURITY?

As mentioned earlier in this chapter, some managers, for a variety of reasons, are reluctant even to discuss the subject of security. In the aftermath of an unsuccessful attempt to burglarize a bank vault, a bank operations manager learned that the anti-intrusion system was 10 years old and somewhat antiquated. "Why didn't the alarm company keep me advised of the necessity to upgrade my system as the state of the art improved?" he asked. The answer he received was, "The alarm system functioned adequately for 10 years with no problem. Would you or the bank have authorized an expenditure of several thousand dollars to upgrade the alarm system before you had this attempted burglary?" The bank representative reluctantly admitted that he would not have. This case is typical of what we call the "knee-jerk reaction" to security. One security consultant with whom the author is acquainted describes it as locking the barn after the horse has been stolen. We all know that the only thing this protects is what the horse left behind.

Nevertheless, it is not uncommon to find that management's attention is obtained only after a serious problem, one pointing to the lack of adequate protection, is brought to its attention. The first reaction seen by management is often one of overkill; the response pendulum swings from complacency to paranoia, when the facts indicate that a proper response should be somewhere in between. Given this situation, an unscrupulous alarm salesman may be presented with an opportunity to prescribe an alarm system worthy of consideration by the manager of the bullion vault at Fort Knox. The true professional will, as the first order of business, try to bring the situation into perspective by calming the fears of the clients.

There are some things a security director or consultant can do to convince top corporate management that proper security is worth spending some money to obtain. Some methods that have proved successful are the following:

1. Establish a meaningful dialogue with the decision makers in the management hierarchy. First, try to ascertain their feelings about

security. What do they really want a security program to accomplish, if in fact they want anything? Do not be surprised to learn that some management personnel regard security as a necessary evil and thus worthy of little attention (that is, time, money, or other resources). Marshal the facts. Research the history of security losses experienced by the company. Use this information to develop trends and projections.

2. When collecting data to support your position, deal in principles, not personalities. Use the technique of nonattribution for all unpublished sources of information. With published sources, such as interoffice memos, extract the pertinent data if possible. Avoid internecine power struggles at all costs. Maintain a position of objective neutrality.

3. Be as professional about security as you can. The better you are at your job, the greater attention you will command from your supervisors. There are many avenues you can explore to develop the information you need, such as developing contacts with other security professionals who share similar problems. Don't reinvent the wheel: attend security seminars, purchase relevant books, study, and do research. Stay current.

4. When making a proposal to management, hit the highlights, and make your proposal as brief as possible. Save the details for later. In any proposal that will cost money, make certain you have developed the cost figures as accurately as possible. If the figures are estimated, label them as such, and err on the high side.

5. It is a wise man or woman who knows his or her own limitations. If you need outside help (and who doesn't from time to time?), do not be reluctant to admit it. Such areas as electronics, advanced CCTV, and sophisticated anti-intrusion alarm systems are usually beyond the capabilities of the security generalist. Do some studying. Know where to go to get the help you need.

6. Suggest that management hire an outside consultant. Competent security professionals have nothing to fear from a second opinion. Often, the "expert from afar" has greater persuasiveness over management than do members of their own staff. More often than not, the consultant will reinforce your position by reaching the same conclusions you did and making the same or similar recommendations.

7. Present your position at the right time. Recognize that management's priorities are first and foremost the generation of profits. To capture management's attention, wait for the right circumstances. It is difficult to predict when these may occur; therefore, have your facts developed and be ready at a moment's notice to make your presentation. It will be too late to do the research when you are called before the board of directors without notice to explain how the breakdown in security that just happened could have occurred, and what you propose to do to solve the problem for the future.

8. Develop a program of public relations. Security represents inconvenience, even under the best of circumstances. Once you have management thinking favorably about your proposal, you will need to sell it to everyone in the organization in order for your ideas to be successfully implemented. Most employees enjoy working in a safe and secure environment. Use this technique to convince employees that the program was designed as much for their safety and security as for the protection of the assets of the corporation.

For a comprehensive treatment of the role a professional security consultant can play and how the consultant can help security professionals properly define their security exposures, refer to Charles A. Sennewald's 2004 text, *Security Consulting*, 3rd edition, published by Elsevier Butterworth-Heinemann. Do your homework in a thorough manner and you cannot help but impress management with your capabilities as a security professional. Remember, be patient. Few have been able to sell 100 percent of their security programs to management the first time out of the starting blocks. And, if you don't remember anything else, remember, as in comedy, timing is everything!

8

Management Audit
Techniques and the
Preliminary Survey

"The objective of a preliminary survey is to quickly and economically determine if a [security] survey (audit) appears to be desirable and is technically and economically feasible."
—Government Accounting Office Audit Guide

AUDIT GUIDE AND PROCEDURES

Audit: Aids to Surveys

Many security professionals, both managers and consultants, found their way to the security field by way of law enforcement, military police, or intelligence activities. Many law enforcement officers have had training as investigators, but few have received training as internal auditors.

There are many similarities between auditing and investigating, but there is one important difference: the audit technique of analyzing facts, drawing conclusions, and making recommendations is absent from the investigator's background and training. Investigators are trained to obtain evidence, report objectively, and scrupulously avoid drawing conclusions or making recommendations. These they leave for their clients. The late J. Edgar Hoover, Director of the Federal Bureau of Investigation, was often quoted as saying, "An investigator must never be wrong in reporting the facts of an investigation. Opinions have no place in an investigative report."

This is not to suggest that investigators do not have opinions or draw conclusions from the results of their efforts. They have and they do, but one will look in vain for these thoughts in the investigator's reports. Investigators are taught not to editorialize but to leave opinions for "the expert witness." Such training, of course, has a real purpose: it is a proven technique for ensuring objectivity and avoiding the trap of partiality in the difficult search for the truth.

The auditor, however, is trained to appraise the truth or falsity of a proposition—not to take things for granted, jump to conclusions, or accept plausible appearance for hard fact. Audit training postulates that accepting appearance for substance is the surest way to arrive at improper conclusions. To be able to differentiate between appearance and substance and draw the proper conclusions is the heart and marrow of the auditor's task.

Both disciplines, to be sure, pursue fact—a fact being something that has actual existence, something that can be inferred with certainty, a proposition that is verifiable. Conjecture, on the other hand, involves propositions carrying insufficient evidence to be regarded as facts. The auditor is trained not only to adduce facts but also to appraise, draw conclusions, and make recommendations from them—the very techniques the investigator is trained to avoid.

In conducting security surveys, one can and should borrow some of the techniques used by internal auditors. It can be said that most security surveys are, in the truest sense, specialized internal audits. In this context, I define *internal (management) auditing* as a comprehensive review, verification, analysis, and appraisal of the various functions (operations) of an organization, as a service to management.

The auditor works with people first and things second, as does the investigator. Both auditing and investigating presuppose a degree of cooperation from the people one encounters along the path to the objective, that path usually being the obtaining of complete information and accurate data, in proper perspective. Successful professionals often obtain this cooperation through the force of their personalities and an empathetic attitude that invariably draws other people out. These individuals, by training or by instinct, understand the art and science of communication—what opens channels and what closes them. But let there be no misunderstanding: both the auditor and the investigator are often feared. Both represent a force and authority that can be a threat to liberty and security, an unknown entity that can adversely affect one's status and well-being, even one's job and livelihood.

The first task of the professional, then, is to allay that fear because fear is a very real impediment to communication, whereas open communication is at the heart of both a successful investigation and a successful audit. There are many techniques that can be used to allay fear, but perhaps the quickest is the use of candor, letting people know, from the beginning, what to expect. It has been said that we often fear most that about which we know the least. Therefore, in the absence of sound justification to the contrary, there should be complete candor and rapport with everyone with whom we come in contact from the outset of the survey or audit until its conclusion. This brings us to some of the methods used by auditors in conducting successful surveys.

Fieldwork

Perhaps as much as 50 percent of survey work is done in the field. The other 50 percent is usually divided equally between planning the survey (logistics) and analyzing the data before writing the survey report. However, some security consultants report that they often spend as much or more time compiling data and analyzing it, and then writing the report, as they did conducting the fieldwork portion of the survey. This is especially true in complicated reviews.

Fieldwork, as considered here, consists of collecting, data, records, and procedures wherever they are found. This process may have an effect on the operation and thus on the results of the survey. Reduced to its simplest terms, fieldwork is largely measurement and evaluation of the effectiveness of the security program (or the lack of one) under review. To be meaningful, measurement must have as its basis an objective standard. By standard, I mean a level of acceptability against which things measured can be compared. Each part of the survey must be approached with the thought that it can be effective if it determines quality in terms that can be objectively measured and compared with an acceptable practice or standard, if one exists.

Only by using these criteria can one measure intelligently and with objectivity. When surveyors cannot measure, they must use extreme caution because in such a case they can produce only subjective observations, not objective reports based on recognized standards or accepted practices. When no standards or acceptable practices exist, obviously the surveyor must construct them. Likewise, when only technical standards exist, measurements obtained must be validated by one who is technically qualified to render a judgment. As surveyors apply recognized standards or acceptable practices, they must not hesitate to evaluate them to determine whether they are valid and not obsolete. It is no understatement to say that without standards or acceptable practices there can be no meaningful measurement; without measurement, in turn, fieldwork becomes conjecture and not fact. Remember the saying, "If it can be measured, it's a fact; otherwise it's an opinion."

There are many methods and approaches to fieldwork, and the one selected often depends on the individual approach of the person making the survey. Nevertheless, fieldwork usually takes the form of observing, questioning, analyzing, verifying, investigating, and evaluating, though not necessarily in that order.

Measurements normally concern at least three aspects of the security operation: quality, reliability, and cost. (Obviously more than three aspects can be programmed into the survey, depending on the results desired.) Regardless of how many aspects of the operation are being reviewed, the primary questions become, Is the procedure, technique, hardware, or electronic device being used effective? Does it properly address and solve the problem, with a known degree of reliability? Does it perform at less cost than others that might be just as (or more) effective with respect to the problems faced and the overall objective or results desired?

The objective of measurement is to assess the adequacy, effectiveness, and efficiency of an existing or proposed system. This is accomplished by applying one or more of the six basic methods of fieldwork that follow.

Observing

Observing is seeing, not just noticing or looking. As Arthur Conan Doyle's Sherlock Holmes was fond of saying, "You see, Watson, but you do not observe." Observing implies a careful knowledgeable look at people and things and how they relate one to the other. It is a visual examination with a purpose, a mental comparison with practices and standards. The ability to observe and evaluate comes from experience. The broader one's experience, the better one observes, and the more alert one is to deviations from the norm.

Questioning

Questioning during a survey occurs at every stage of the proceeding. It may be in oral or written form. Oral questions, of course, are the most common and at times the most difficult to pose. To get the truth without upsetting people during the course of a survey is not an easy assignment. If subjects detect an attitude of cross-examination or an inquisitorial tone, they may promptly raise their defenses, which become barriers to communication. The results may be wrong or incomplete answers or, worse, no answers at all. Most successful practitioners have developed the interview technique into an art. They use this, perhaps the most common tool of our trade, with a high degree of effectiveness. Remember, there is a big difference between an interview and an interrogation. An interview presupposes voluntary participation.

Analyzing

Analyzing is nothing more than a detailed examination of a complex entity to determine the true nature of its individual parts. It presupposes intent to discover hidden qualities, causes, effects, motives, possibilities, and probabilities. In contrast, if one examines an operation as a whole, one cannot perceive the intricate relationships of the diverse and varied elements that make up a complex function or an unusually large activity. Any composite, no matter how large or complex, can be analyzed by division, by breaking it down into its separate elements and then observing trends, making comparisons, and isolating aberrant transactions and conditions. Frankly, the job can be done no other way.

Verifying

One verifies by attesting to the truth, accuracy, genuineness, or validity of the matter under scrutiny or inquiry. This implies a deliberate effort to establish the accuracy or truth of some affirmation, by putting it to the test. An example is a comparison with other ascertainable facts, with an original, or perhaps with another acceptable practice or standard. Verification may also include corroboration—the statement of another person or a validation by objective practices or standards, usually found elsewhere.

Investigating

An investigation is an inquiry that has as its aim the uncovering of facts and the obtaining of evidence to establish the truth. An investigation may occur as a part, or as the result, of a survey, but it is not restricted to some impropriety. If it is in fact to be restricted to an impropriety, it must be divorced immediately from the survey and referred to the proper authorities within the entity being surveyed for appropriate handling.

It is not unusual in the course of a security survey to uncover suspected fraud. Colleagues of mine and I did just that in the course of a security survey at a large hospital and medical complex. The receiving (intake) office had a thief, who had been stealing patients' valuables; the discovery necessitated a separate inquiry, which proved successful and produced a conviction. When uncovered, fraud and other forms of employee malfeasance must be dealt with outside of the scope of the review. (See Appendix B, Danger Signs of Fraud, Embezzlement, and Theft.)

Evaluating

To evaluate is to estimate worth by arriving at a judgment. It is to weigh what has been analyzed and determine its adequacy, effectiveness, and efficiency. It is one step beyond opinion, in that it represents the conclusions drawn from accumulated facts. Evaluation, by necessity, implies professional judgment. Professional judgment is a thread that runs through the entire fabric of the security survey.

Evaluation in a survey occurs constantly throughout the duration of the project. In the beginning, one must determine which programs and procedures will be reviewed, which processes and operations are to be tested, and how big a sample must be obtained for the test to achieve the degree of sample reliability needed. Finally, as the results of the survey accumulate, one must evaluate what the results imply—fact finding without evaluation is a clerical, not a professional, function.

Evaluation obviously calls for judgment, and this is perhaps the real test of the true professional. There are bound to be questions and imponderables. If you find none, it may be time to leave well enough alone and get out of the game. The true art is in recognizing where the questions lie. Once we identify the problem, we are well along the way toward the solution. Like a doctor making a diagnosis, only by properly identifying the problem can the security professional adequately prescribe the solution.

Mature professionals evaluate results almost intuitively, and more often than not they are correct. We of lesser experience can benefit from a more structured, formal, and organized approach to the evaluation of the results obtained. For example, in evaluating a failure to meet specific standards or acceptable practices, one might ask the following questions:

- How significant are these deviations?
- Have they or will they prevent the operation from achieving its objective or mission?
- Who or what can be hurt or injured?
- Could the injury perhaps be fatal to the enterprise?
- If corrective action is not taken, would the deviation be likely to recur?
- How did the problem surface in the first place?
- What were the causes? What event or combination of events caused the failure to occur?
- Will the event or combination of events cause the observed results or failure every time?

Obviously, to recommend corrective action, we must first answer some of the above basic questions. In following these procedures, surveyors find themselves in a constant posture of evaluating. One must constantly query everything under review, with questions such as the following:

- What is the real problem? (Not necessarily what management thinks it is—what is it really?)
- What are the relevant facts? What are the processes, systems, procedures, policies, and organizational structures? What were they in the past? What will they be in the future?
- What is presently being done about this problem at other locations within the company? At other companies?
- What are the causes? The number and variety of causes? The root cause as well as the surface cause? When and how do these causes affect the overall problem?

- What are the possible solutions, the alternatives, the cost, and the answers to the problem? What are the possible side effects, advantages, or disadvantages to the proposed solution?

Only by constant probing, questioning, analyzing, and evaluating do surveyors uncover system and performance defects, establish reliability, and develop cost-effective solutions to their clients' problems.

THE PRELIMINARY SURVEY

Definition and Purpose

The basic purpose of the preliminary survey is familiarization, based on more than a mere discussion or brief observation of the tasks to be reviewed. It presupposes an ability to perceive the true objectives of the operation and to locate and evaluate the key control points, if any exist. Also, one must understand the management concepts being used and the qualifications and abilities of the employees responsible for the success of the operation. (See Box 8.1 for a sample statement of purpose.) A properly planned and implemented preliminary survey allows one to develop a well-thought-out program, to deploy one's efforts efficiently and economically, and to form a firm foundation for the detailed examination that will follow.

Although a poor preliminary survey (or none at all) can easily result in a poor audit, a good preliminary survey can ensure an intelligent examination and may also substitute for many parts of the final examination. It is the simplest way to cut through the mass of detail that often obscures the objective and to get the job started quickly and on the right track. Another advantage of the preliminary survey is to test the client's sincerity and avoid misunderstandings regarding the client's expectations from the project at the outset.

Many security and safety operations in large companies are extremely complex. The immediate task becomes one not only of identifying and understanding these operations but also of analyzing and evaluating them and recommending improvements designed to accomplish the job with greater efficiency and at less cost. Admittedly, this is no easy task in a large and complex industrial, research and development, or military environment. This is especially so when one adds to the already identified internal complexity of the operation such factors as the external environment (community), legal restrictions, ecological and environmental protection procedures, public relations, employee relations, union activities, government regulations and restrictions (especially the Occupational Safety and Health Administration, or OSHA), and stockholder interest. Additionally, since the events of 9/11, we must also consider the prospect of international terrorism, as it may (or may not) affect the operation.

As complicated as the assignment may appear initially, it is well to remember that there is no mass of data so large, and no operation so complex, that it cannot be given a semblance of order and then be arrayed, evaluated, and summarized in a logical, organized, methodical manner.

A preliminary survey should answer at least the following questions:

- What is the operation?
- Who or what is responsible for the operation?

- Why is it done (where and when)?
- How is the operation accomplished?

Box 8.1. Preliminary Survey—Statement of Purpose

Attention: Name of Client
Reference: University of _____

The preliminary survey will be a primary overview of the university to identify major problem areas within the system affecting security. This initial review will become the basis for defining the parameters of the final survey. It will include the following:

- Interviews of officials concerned with administration of revenue-producing and accounting programs to identify loss exposure.
- A physical tour of selected campus locations to familiarize consultant with the revenue-generating operations of the university.
- Preliminary interviews with selected on-site operations personnel for basic orientation with cash-handling methods and control procedures.

The agreed upon scope of the final survey will emphasize internal control procedures, with physical security and emergency planning secondary. A complete evaluation will include specific recommendations for appropriate administrative controls, hardware application, and personnel to complement the system already in use to achieve effective cost-control security.

We will also provide downstream inspection to ensure that agreed upon recommendations are being properly implemented.

Sincerely,
James F. Broder, CFE, CPP, FACFE, Security Consultant
PMG

In short, get what is important, and get it with a minimum of delay. Focus on the highlights and forget, for the moment, the details; they will come later. However, don't let haste interfere with order and methodology. To go into the program's initial operation without a well-prepared agenda may well leave the client with the impression of disorganization, the exact opposite of what one would hope to instill at this (or any other) stage of the security review.

Security Consultant Charles A. Sennewald, CPP and I were contacted and asked to do a physical security survey of three large petrochemical plants near Bombay, India. The client asked for a brief proposal outlining our plan to complete this complex project. We recommended first doing a preliminary survey and provided an estimated cost (travel, time, and expenses) to do the survey work and to write the preliminary survey report. Our planned approach was well thought out; the preliminary survey was estimated to take about 3 days (a total of 24 hours) of consulting time. Upon submission, however, our bid was rejected. The client had expected us to absorb the cost associated with the preliminary survey, in return for the possibility of being selected (maybe) to do the principal project, which would have taken about 6 months to complete.

We later learned that this was the customary way some people in that country did business. Notwithstanding, the proposal we sent included a comprehensive outline for performing a preliminary survey. The project was so large and decentralized that it would have been impossible to accomplish the principal task without first performing a preliminary survey.

The Initial Interview

At the initial interview (or "opening conference" in audit parlance), the nature of the questions asked will vary depending on whether the survey is organizational, functional, or operational. If it is organizational, people-oriented questions will be the general rule. Functional surveys are more concerned with the actual work flow, which more often than not crosses organizational lines. In operational surveys, one is primarily interested in hardware and, equally important, in the software and related procedures concerned with the effective use of the hardware. Regardless of the type of survey, the initial interview should elicit answers to the following questions: What does management perceive to be the major problem areas? What does management hope the survey will accomplish in regard to solving these problems?

At the initial interview, it is wise to have prepared in final form a document called the management memorandum. This memorandum, to be signed by the highest authority possible, introduces the survey team, describes the objective of the survey, solicits the assistance and cooperation of all company employees, and authorizes access to all documents and information that may be requested in the course of the survey. Copies of this signed memorandum, when possible, should be sent to affected department heads in advance of the arrival of the survey team. In this regard, it may be helpful for the people doing the survey to be introduced at a meeting of all department heads.

At this meeting, the management memorandum can be read and distributed, and any questions regarding operational authority can be answered. At a minimum, someone from the company must be assigned to escort and introduce members of the survey team to department heads and employees with whom they will be working. We prefer, however, the department head meeting because such a meeting can go a long way toward establishing rapport and cooperation. It also gives the department heads directly involved a chance to ask questions and receive answers that ideally will allay any fears or apprehensions they brought with them to the meeting. At the very least, it serves as a vehicle to get them involved in the survey program at the outset.

After the preliminary survey has been completed, it is useful to have a second meeting with the client. The purpose of this second meeting is fourfold:

1. To give the client a brief report of initial impressions obtained
2. To explain how the surveyor perceives the objectives, activities, or functions under review

3. To establish a meeting of the minds as to just what is to be accomplished
4. To briefly outline to the client the general plan of attack

It is essential that both parties to the survey be in total agreement with regard to the objectives of the project. In this way, any misunderstandings can be resolved so that the principal job can proceed without delay.

Obtaining Information

What Information to Obtain

The preliminary survey, and in fact the primary survey as well, will move along rapidly and systematically only if one has a clear idea of what data are needed and where to find them. Some, but by no means all, of the basic sources one should consider are as follows:

The Charter for the Organization.   Copies of policy statements, directives, statements of functions, responsibilities, goals, and delegations of authority will be needed for review. In addition, one will need job descriptions of the people directly involved in the activity, if available.

Beyond the written word, it is essential to focus on the objectives of the operation— what is its real mission (not necessarily what the official statements say it is). It is not uncommon to find that official job descriptions are mere window dressing, or that they have not kept pace with changing times and aims. The organization of the operation may include the following:

- Organizational charts
- Position descriptions of the operation in the overall company structure
- The nature, size, and location of ancillary or satellite activities
- Interfacing operations and their relationship to the activity under review (safety, for instance, when the primary review is security)

Financial Information.   One will want to obtain for review all financial data that have a bearing on the subject under scrutiny, either directly or indirectly. Obtain copies of the company's annual reports, if available.

Operating Instructions.   It is essential to obtain an accurate picture of the flow of records and other data. One of the simplest ways is by flowcharting the activity. Flowcharts (discussed later in this chapter) can provide a useful picture of the operation and at the same time highlight gaps and duplications in procedures, as well as pinpoint risk areas for later scrutiny.

Problem Areas.   During the entire survey, one should keep in mind the problems mentioned by management at the preliminary (opening) conference, as well as any deficiencies found in prior surveys, audits, or reviews. Also, focus attention on the procedures or controls that have supposedly been designed to alleviate the difficulties and problems and at the same time reduce the risks.

Matters of Special Interest. One will be especially interested in exploring any new areas mentioned during discussions with management that were of concern to them during the preliminary (opening) conference.

Sources of Information

Some, but certainly not all, of the possible sources of information available during the survey are as follows:

1. Discussions with supervisors and employees directly engaged in the activity under review. One cannot overemphasize the importance of these people, because:
   a. Not only are they usually aware of the problems, they often have worked out the solutions as well.
   b. It is essential to obtain their cooperation because in the final analysis, these are the people who will be responsible for implementing many of the recommendations made as a result of the survey.
   c. If they feel they have played a real part in the development of solutions to the problem, they will be more inclined to work for the success of the recommendations. The reverse is, unfortunately, also true.
2. Discussions with supervisors downstream and upstream of the operations under review
3. Correspondence files
4. Prior survey, audit, or inspection reports
5. Incident and crime reports
6. Budget data
7. Mission or objective statements or reports
8. Procedural (operational) manuals
9. Reports by or to government agencies, both state and federal (OSHA, for example)

Physical Observation

Observations or inspections should be conducted in two phases. The first is a familiarization tour of the entire facility to obtain "the big picture." At this time, the various departments are identified, and the managers and supervisors introduced previously are seen in their normal work environments.

Notes are made, but few questions are asked at this point. In a small operation, one tour may be sufficient to accomplish the desired objective. In a large, complex operation, it may be necessary to make a second or even a third tour before feeling comfortable with the facility and its environment. The second (or subsequent) tour may be made in connection with flowcharting various parts of the activity or operation.

Flowcharting

Flowcharting is an art that with proper practice can become an invaluable survey tool. Making a flowchart is the easiest way to obtain a visual grasp of a system or procedure, and it is a ready means of analyzing complex operations that cannot easily be reduced to meaningful narrative description. Figure 8.1 shows some standard flowchart symbols and a legend describing each symbol used. Sometimes a simple sketch may suffice; at other times, it may be necessary to use plastic overlays to describe detailed and complex operations. Figure 8.2 is an example of a formal flowchart describing a complex operation. Flowcharts, however, need not be formal or greatly detailed to accomplish the task. Table 8.1 is an example of a simple, informal flowchart.

Symbol	*Explanation*
	Starting point in the flow of documents
	Document
	Direction in the flow of a document
	Control point
	Direction in the flow of information
	Permanent file of documents, alphabetically, numerically, and by date
	Temporary file of documents
	Operation or action
	Document destroyed
	Document signed
	Document initialed
	Punch card
	Report of computer printout
	Book or ledger
	Source of posting to general ledger

FIGURE 8.1 Standard Flowchart Symbols.

SUMMARY

Fieldwork is measurement—it is measuring "what is" against "what should be." This requires both the methods of measurement and the existence of acceptable practices and standards for comparison. Security surveys are usually concerned with measuring at least three basic factors: quality, reliability, and cost.

To measure, one must go out into the field and perform surveys or, if a program already exists, to review and to test it for reliability. Fieldwork is performed using the techniques of observing, questioning, analyzing, verifying, investigating, and evaluating. The largest portion of fieldwork is gathering data and accumulating evidence, which must then be analyzed and evaluated before recommendations can be developed. It is in the proper evaluation of the data obtained that true professionals meet their test.

FIGURE 8.2 Formal Flowchart.

ACCOUNTS PAYABLE
DEPARTMENT

DATA PROCESSING
DEPARTMENT

Controls:
Stamps controlled by registers in Purchasing Department. Only authorized depart-
ments and personnel may validate RTPs or use DD, system.
Unvalidated requests are returned to requesters.
Improperly approved P/S's challenged.
Payments not made without proof of receipt.
Invoices not required.
Record of commitments helps establish cash needs.
Check signed in Finance Department (not on chart) where it is mailed to supplier.

⟨7⟩ Reports to cognizant managers on unearned discounts monitors timeli-
     ness of processing P.O.'s, P/S's, RMs and payments to supplier.

⟨8⟩ Reports warn that D.D. privileges may be withdrawn.

Abbreviations:
D.D. – Direct Delivery
P.O. – Purchase Order
P/S – Packing Slip
RM – Receiving Memo
RTD – Request to Purchase

FIGURE 8.2 (*Continued*)

The preliminary survey charts the course for the main voyage. It often provides a clear enough view of certain activities to eliminate the need for further review of these operations. The time spent on the preliminary survey is well spent. Done properly, it will ensure a more efficient and economical final review.

In the preliminary survey, one gets to know people, understand operations, and focus on objectives, controls, and risks. One is then in a better position to perform the main survey in an intelligent, effective, and efficient manner. The preliminary survey is the road map, essential for a long and difficult journey.

Table 8.1 Informal Flowchart: Procurement of Materials

Purchasing Department	Receiving Department			A/C Payable	Inspection	Stores
	Office	Dock	Hold Area			
Purchase orders and changes are prepared and sent to: 1. Supplier 2. Accounts Payable 3. Receiving Department 4. Buyer 5. Purchasing files	The master P.O. is held in temporary files awaiting receipt of materials and shipping notice. Upon receipt of shipping notice, the receiving information is added to the ditto master of the P.O. to create the Receiving Memo.	Materials and shipping notice are received. The S/N is sent to Receiving Office. The materials are sent to the hold area.	Materials are held until the Receiving Memo is prepared. Thereupon the materials are sent to Inspection.	Evidence of receipt is matched with copy of P.O. No invoice is required. If match is satisfactory, payment to supplier is approved.	Material is inspected. Unsatisfactory material is sent to hold area. Satisfactory material is sent to stores.	Materials are stored awaiting requisitions from using departments.

9

⬜⬜⬜
⬜⬜⬜
⬜⬜⬜

The Survey Report

"Did you ever stop to think how odd it is that you have to learn how to write your own language? Why? What is there to learn?"
—Rudolf Flesch and A. H. Lass, *The Way to Write*, Harper & Row, 1963.

There is little likelihood that the average security professional was born with the writing mastery of, say, Ernest Hemingway. Writing is an art, and like most arts, it must be constantly practiced if it is to become a useful and natural talent. Although few of us can learn to write with the flow and style of a great master, we can, with effort and practice, greatly improve our ability to communicate by use of the written word.

Many of us find fieldwork the most exciting and challenging part of our daily assignments. We all know many professionals who are extremely competent investigators and auditors but who, when it comes to writing reports, leave much to be desired. Good writing requires good thinking. If one's concepts are confused and tangled, one's reports will reflect the same problems. If one's thoughts are muddy and don't seem to establish a bridge between cause and effect, the resulting written report is bound to reflect this confusion. Unfortunately, we are likely to be judged more often by our ability to write a good report than by our ability to do good fieldwork. Our professional ability and efficiency will be demonstrated before the eyes of many by clear, concise, complete, and accurate reporting.

I need cite but one example to make my point. We were contracted to do an investigation/audit regarding a possible kickback scheme between a vendor and a staff employee of a public agency in a large city of southern California. When we submitted our final written report to the client, we expected it to be reviewed by no more than five senior management officials. Instead, the report was reproduced, and copies were presented to the entire board of directors. Also, copies were furnished to the district attorney's office for review and possible submission to the grand jury. And, eventually, a copy of our report was leaked to the press! Many people, most of whom the writer never met, were thus in a position to judge the quality of our product by merely reading the report. The message here is, "Every report you write bears your name. Write one bad or inaccurate report, and it will haunt you forever."

"I Must Write, Therefore I Shall"

What makes good writing? The answer seems to lie in good fieldwork, a well-structured outline, copious notes or working papers, and dogged persistence. If one has done an adequate job in the field, one's notes and working papers will be full of facts and figures. Only then can one be sure that adequate material exists for a good report. At this point, it is well to remember the Chinese proverb, "A journey of a thousand miles begins with the first step." Sit down and start writing. The first order of business will be to prepare an outline. An example of an effective reporting outline can be found in Box 9.1.

Box 9.1. Security Survey Report

I. Purpose. State the reason for conducting the survey. The purpose could encompass all of the subject matter in this outline or be restricted to a specific portion or spot problem; it could encompass one location or cover the entire corporation.

II. Scope. Describe briefly the scope of the survey effort. What was actually done, or not done, in some cases? For example:
   A. People interviewed
   B. Premises visited, times, and so forth
   C. Categories of documents reviewed

III. Findings. This section includes findings appropriate to the purpose of the survey. For example, a survey report on physical security only would not contain findings related to purchasing, inventories, or conflicts of interest unless they have a direct impact on the physical security situation.
   A. General: Provide brief descriptions of facilities, environment, operations, products or services, and schedules.
   B. Organization: Provide brief description of the organizational structure and number of employees. Include details regarding the number and categories of personnel who perform security-related duties.
   C. Physical security features: Describe lock-and-key controls, lock hardware on exit and entrance doors, doors, fences, gates, and other structural and natural barriers to access and entry. Describe access controls, including identification systems and control of identification media; the security effect of lighting; the location, type, class, installation, and monitoring of intrusion alarm and surveillance systems; guard operations; and so forth.
   D. Internal controls: Describe methods and procedures governing inventory control, shrinkage, and adjustments; identification

and control of capital assets; receiving and shipping accounts receivable; purchasing and accounts payable; personnel selection; payroll; cash control and protection; sales to employees; frequency and scope of audits; division of responsibilities involving fiduciary actions; policy and practices regarding conflicts of interest; and so forth.

E. Data systems and records: Describe physical features and procedures for protecting the data center, access to electronic data processing (EDP) equipment, the media, and the data; types of application systems in use; dependency; backup media and computer capacity; and capability for auditing through the computer. Identify essential records and how they are protected and stored.

F. Emergency planning: Describe status and extent of planning, organization, and training to react to accidents and emergencies, such as emergencies arising from natural disasters, bombs and bomb threats, kidnapping and hostage situations, and disorders or riots. Consider these in three phases: pre-emergency preparations, actions during emergencies, and post-emergency or recovery actions. Discuss coordination with and utilization of external resources, for example, other firms or government agencies.

G. Proprietary information and trade secrets: Discuss the extent to which these categories of information are recognized and how they are classified; means of protection, for example, clearance, accountability, storage, declassification, disposal; and secrecy arrangements with employees, suppliers, and so forth.

IV. Conclusions. Evaluate the protective measures discussed in part III, Findings. Identify specific vulnerabilities and rate them as to seriousness, for example, slight, moderate, or serious, or similar terms of comparison.

V. Recommendations. Using the systems design technique, the consultant develops specific recommendations for appropriate applications of hardware, administrative controls, and "person power" that will complement the protective measures already in use to provide effective controls of the vulnerabilities identified in part IV, Conclusions. Cost/benefit ratio considerations are applied to each recommendation and to the structure as a whole.

Note: The client's experience is always reflected in the appropriate sections of the report.
Source: Compliments of Charles E. Hayden, PE, CPP, Assistant Vice-President and Senior Security Consultant (Retired), M & M Protection Consultants, San Francisco, California.

If you sit around and wait for divine inspiration to give you the perfect beginning, you may well wait forever. You have to be able to develop a discipline that says, "I must write, therefore I shall," and immediately take pen to paper. It will matter little that first passages of the draft are poorly phrased or could be worded better. Doubtless, the final report will have been reshaped and revised a number of times. The rule to remember is: "There is no such thing as good writing; there is only good rewriting." Only after painstakingly preparing dozens and dozens of reports do one's writing skills improve. Only then does the job become less painful and the product more professional. Finally, when writing truly becomes a joy, professional reports will follow. Few of us, however, will reach this exalted plateau, and it is not the purpose of this text to chart a course to becoming a professional writer. It is our purpose, however, to describe some of the time-tested methods for writing better reports, thus increasing one's ability to better serve clients and management.

Survey reports generally have two functions: first, to communicate, and second, to persuade. The findings, conclusions, and recommendations in the report are very important to management. For the report to communicate effectively, the channels must be clear, the medium must be incisive, and the details must be easily understood. Also, the story must be worthy of the material. Too often the skill and effort expended during the fieldwork portion of the survey are lost in the murky waters of a poorly written report. A dull or poorly written report will not penetrate the upper circles where the story needs to be told and where the persuasion needs to occur. Write your report expecting it to be read by the top decision maker in the organization, and you will always be safe.

FIVE CRITERIA OF GOOD REPORTING

The ability to write good reports can be developed. With the proper desire, the right amount of effort, the right standards, and the proper techniques, much can be accomplished. I stated earlier that fieldwork is largely measurement, which implies the existence of acceptable practices and standards (criteria). Reports can also be measured against standards. A good report must meet the following criteria: accuracy, clarity, conciseness, timeliness, and slant (or pitch). To assist the reader in improving one's ability to write reports, we will consider each of these criteria separately.

Accuracy

A reader should be able to rely on the survey report, because of its documented fact and inescapable logic. Additionally, the report must be completely and scrupulously factual, based entirely on the evidence at hand. Likewise, the report should speak with authority. It should be written and documented so as to command belief and convey reliability. Facts and figures, as well as statements and recommendations, must be supported by the evidence. Statements of fact must carry the assurance that the person doing the survey personally observed or otherwise validated them. When it is necessary to report matters not personally observed, the report should clearly identify the source.

The writer must be careful to avoid personal attribution unless, of course, one can personally certify the existence or extent of the condition reported.

Another important facet of accuracy—and one often overlooked—is perspective, the reporting of facts in the proper light. As an example, it would be reporting out of perspective to show how one of a dozen related activities may be deficient without showing how all of the activities relate, one to another, in order of importance. This is sometimes referred to as "balance."

For a report to be accurate then, it must encompass truth, relevancy, and perspective. It must also have balance.

Clarity

One cannot write clearly about subject matter one does not understand. Clarity implies communicating from one mind to another, with few obstacles in between. To accomplish this, the report writer must have a firm grasp of the material at hand. Until writers reach this level of understanding, they cannot take pen in hand and start to write a report. They would be better off returning to the field and continuing their activities until they feel totally comfortable with the material obtained.

Poor structure is an impediment to clarity. An orderly progression of ideas lends to clarity and thus understanding. Therefore, as most professional writers have found, one of the best aids to effective written communication is the prepared outline, as shown in Box 9.1.

Acronyms, abbreviations, and technical jargon should be avoided when possible. If it is not possible to avoid their use, then at least give the reader at the outset an explanation of the term or initials. For example, at first mention: "Department of Defense (DOD) Bulletin #12 states . . ." not "DOD Bulletin #12 states . . . ." Do not assume that everyone reading your report will be familiar with a term, acronym (initials), or technical jargon or slang terms common to the environment being studied. Quite to the contrary, good writers assume that their audiences are not familiar with these oddities and take the time to explain them in order to enhance clarity and understanding.

Likewise, reporting a finding without properly setting the stage can lead to misunderstanding. Only by reporting relevant information and background can the author expect the reader to understand the process or condition and thus appreciate the significance of the finding. If one is recommending a new procedure, one should first state what procedure, if any, now exists and why it isn't working. This makes the reader fully cognizant of the procedure in question and much better positioned to consider the proposed change favorably.

Long discussions of technical procedures muddy the waters. Here, the liberal use of flowcharts, schedules, exhibits, and graphs can be a real aid to clarity and ease of understanding. Be ever mindful of the old Chinese proverb, "One picture is worth a thousand words." Such aids benefit both the reader and the writer of the report. If it is axiomatic that one of the purposes of the survey report is to stimulate action, along the road to change, then the report must be an effective stimulant. To be effective, the writing in the report must be clear!

Conciseness

To be concise means to eliminate that which is unessential. Conciseness, however, does not necessarily mean brevity. The subject matter may dictate expanded coverage. Conciseness does mean elimination of that which is superfluous, redundant, or unnecessary. In short, it is the deletion of all words, sentences, and paragraphs that do not directly relate to the subject matter in question. This is not to suggest an arbitrary, telegraphic style of writing, though in some instances this style may be appropriate. It is, however, to suggest that easy flow and continuity of thought do not depend on excessive verbiage. Short, simple, easy-to-understand sentences are the general rule. Whenever one comes head-on into a long, complicated sentence, continuity of thought on the part of the reader generally is replaced by confusion. Long sentences should be dissected and rewritten to obtain a comfortable, happy medium.

Do not, however, confuse conciseness with lack of data. There must be sufficient detail in the report to make it meaningful to all levels of the audience. Those involved in the intimate day-to-day operations, as well as the president and chairman of the board of directors, must have sufficient information to understand the problems.

Timeliness

Often it takes a while to get management to approve having a security survey or audit initiated. Once the decision is made and the project begins, management waits impatiently for the final result of the survey, the written report. The report should answer all of management's needs for current information. Therefore, the report must be submitted in a timely fashion. One must be careful, though, not be in such a hurry to communicate one's findings that the report is otherwise unacceptable. The survey report must meet all the other criteria for good writing, as well as timeliness.

How do we then meet objectives, timeliness, and thoroughness? One solution may be the interim, or progress, report. When, during the course of a survey, we come across a finding of such importance or magnitude that the details must be reported to management without delay, we can use the interim report. An example may be a serious safety deficiency that if not corrected immediately could lead to injury or death. Another example is if one uncovers the unmistakable danger signs of fraud, theft, embezzlement, or industrial espionage (see Appendix B). Interim reports communicate the need for immediate attention and, one would hope, immediate corrective action. In this regard, one would probably first communicate the deficiency to top management orally, then follow-up immediately with the written interim report.

Interim reports are designed to be short and to the point. They should address one subject, perhaps two at most. These reports should be clearly identified as "interim," and they must contain, usually at the bottom of the page, written disclaimers against accepting them as the final word on the subject being reported—"in the absence of a full inquiry," or words to that effect. The interim report's purpose is to give management an opportunity to focus immediately on the reported condition and get corrective action started without waiting for the final report to be published.

Interim reports notwithstanding, the final survey report should always be submitted in a timely manner.

It is helpful to establish, at the outset of the project, goals and time frames, which include the issuance of the final report. Schedules are necessary to keep survey projects under control. The larger and more complex the project, the more important scheduling becomes. This is especially true for those parts of the survey that are least desirable; report writing unfortunately falls into this category for many people. I have found that once the fieldwork is completed, there is a tendency to delay the reporting part of the job. Nevertheless, the end result of the task is the written report, and management has every right to expect that reports be submitted expeditiously.

Another thought on this subject: I have seen otherwise outstanding fieldwork totally destroyed because a report was not submitted to the client on or before an agreed-upon deadline. A security survey can be an expensive proposition. Clients often become upset, and rightfully so, if they feel they are not getting what they are paying for, in a timely manner.

Slant or Pitch

Finally, the writer should consider the slant or pitch of the report. Obviously, the tone of the report should be courteous. We should also consider the effect it may have on operations personnel. It should strive for impersonality and thus not identify or highlight the mistakes of easily identified individuals, small units, or departments. The report must not be overly concerned with minor details or trivialities. It must avoid sounding narrow-minded, concentrating instead on that which has real meaning and substance. The report should clearly be identified with the needs, desires, and goals of sound management principles. Pettiness must be avoided at all costs.

FORMAT

A report is generally divided into two parts. The first concerns itself with substance— the heart and soul of the project. The second is form, and form usually is dictated by the type of report being prepared, whether it is formal or informal, final or interim, written or oral.

The form of the report can also depend on those to whom the report is being directed. What do prospective readers expect? How much time will they be able to devote to reading the report? The answers to these questions often become the deciding factors regarding the report format used.

Also, different writers employ different formats. There is no universal style or format that can be recommended to satisfy all needs. Novices will usually begin by experimenting until they find a format that best suits their style of delivery. When they find an effective format, they can modify it to meet the specific requirements and tasks at hand. The following elements are usually found in survey reports.

Cover Letter

The cover letter can serve many purposes, not the least of which is to be a transmittal document for the report. The writer may wish to include in the cover letter a brief synopsis of the findings. This would be especially helpful to members of management who are concerned only with the broad overview of the project. The writer must make certain that the details contained in the synopsis are factual extracts from the body of the report. The tendency to take poetic license and "summarize the details" must be carefully avoided, or meaning may become distorted. Whenever possible, cover letters should be limited to one page, for the benefit of the busy executive; also, anything beyond a one-page transmittal begins to invade the province of the attached report.

Body of the Report

Title

The title should fully identify the name and address, with zip code, of the entity being surveyed. It should also include the dates between which the survey was conducted, but not the date the final report is being submitted to management—that goes on the cover letter. Also, the title should clearly state that this is a "SECURITY SURVEY (AUDIT)," in the top center of the first page, in capital letters.

Introduction or Foreword

The introduction should be brief and clear and should invite further attention from the reader. Its purpose is to provide all the data necessary to acquaint the reader with the subject under review. At this point, the writer may wish to identify the sites or departments toured, the people interviewed, and the documents examined. One point of information that is essential here is the authority for the survey; this is usually contained in the opening sentence, for example, "This survey was initiated under the direction of the XYZ corporate management, specifically to identify and evaluate. . . ."

Purpose

The purpose section is a brief description of the objectives of the survey. It must be in sufficient detail to give the reader an understanding of what to expect as a result of the survey. The purpose should be spelled out precisely at this point, so that when the findings are reported, they can be seen to conform to the statements contained in this section. This technique makes it easier for readers to find the substance of any particular item without reviewing the entire report. As an example, in the statement of purpose, both the objectives and findings may be listed in numerical or alphabetical order, for ease in comparison between the two.

Scope

The statement of scope should be a clear delineation of exactly what areas are being reviewed and to what extent they are being examined. It sometimes helps to clarify the issue by specifically spelling out areas that are not covered in a particular survey. This is especially helpful when the title of the report is so broad or general that it may lead the reader to expect much more than the report actually delivers. In a brief report, it may be advantageous to combine the scope and purpose for the sake of brevity.

An example of a statement of scope is, "We limited our review of the purchasing department to those activities that directly relate to sole-source acquisitions. We did not review competitive bid activities in the course of this examination."

Findings

Findings are the product of fieldwork. They are the facts produced by the interviews, examinations, observations, analyses, and investigations. They are the heart and marrow of the survey project. Findings may be positive (favorable) or negative (unfavorable). Findings can depict a satisfactory condition that needs no attention or an unsatisfactory condition deserving of immediate correction.

Positive findings require less space in a report than negative findings. Because they do, however, represent a survey determination, it is essential that they be included. Some writers do not report positive findings; these writers feel, why burden management with matters requiring no corrective action? Others report favorable findings to show objectivity and to give balance to the report. I favor reporting both positive and negative findings so that management can see the operation under review in total perspective.

The statement of reported findings usually includes a summary of the findings and the criteria or standards of measurement used. Also included are the conditions found, and, for deficiencies, their significance, causes and effects, and recommendations for corrective action.

In preparing to present a negative finding, one should be able to answer the following questions:

- What is the problem?
- What are people (procedures) supposed to be doing about this problem?
- What are they actually doing, if anything?
- How was this situation allowed to happen?
- What should be done about it?
- Who is responsible to ensure that corrective action will be taken?
- What corrective action is necessary to remove the deficiency?

Using these questions and other obvious questions as criteria, the writer can be satisfied that he or she has done an adequate job, in a fair and impartial manner. Only then can the report writer set forth the findings in a way that will satisfy the client that the recommendations have merit and meaning. It is well to remember, however, that the sole responsibility to initiate action, based on the recommendations, is the duty of the person who receives the survey report. The security consultant can also assist the client in the implementation phase of the recommendations, to the extent desired by the client.

Statement of Opinion (Conclusions)

Not all reports contain the opinions of or conclusions reached by the members of the survey team. Those that do usually provide a capsule comment that reflects the professional opinion and judgment of the surveyor regarding those activities that have been reviewed. When they are reported, comments can be both positive and negative.

Opinions are nothing more than the professional judgments of the people making the review. Some professionals believe that management is entitled to receive and review their opinions and conclusions. Some believe otherwise, eliminating conclusions entirely from the report and moving instead directly to the findings and recommendations.

Regardless of the report writing technique used, when one sets forth a conclusion or an opinion, it must be supported by fact and fully justified. Also, the opinion must be responsive to the purpose of the review as set forth in an appropriate part of the report, or it may well be superfluous.

Although it goes without saying that opinion more often than not finds fault, one should not hesitate to express positive opinions when compliments are deserved. This should be done not only in the interest of fairness but also to lend balance to the report. An example of an opinion (conclusion) is, "In our opinion, the procedures designed to ensure an orderly evacuation of bedridden patients in the event of fire or natural disaster were found to be adequate given the obvious limitations of a high-rise environment at this hospital." Or, "In our opinion, the transit system was determined to be safe and secure based on our analysis of the Reports of Serious Incidents made available for our review."

SUMMARY

Reports are evaluated according to their accuracy, clarity, conciseness, timeliness, and slant (or pitch). Whatever format is used (and formats come in many styles), it should include one or more of these common elements: a cover letter and the body of the report, which includes a title, a foreword or introduction, the purpose, the scope, the findings, the opinions, the conclusions, and the recommendations. Good writing takes constant practice and effort, but any writing can be improved by using certain techniques and adhering to established criteria or standards. Begin by preparing a logical outline, as the skeleton to which the muscle can be added later. A liberal use of flowcharts, schedules, exhibits, and graphs can be a real aid to clarity and understanding. Finally, remember the golden rule of report writing: "There is no such thing as good report writing, there is only good rewriting."

10

Crime Prediction

Eugene Tucker, CPP, CFE, CBCP
Contributing author

"What will amount to adequate security varies in each individual case. No case has determined how many guards might be required or provided any other specifics as to what constitutes adequate/reasonable security."
—*Lopez v. McDonald's*, 193 Cal. App. 3d 495 (1987)

This chapter examines crime as an element of hazard identification and provides some guidelines to predicting its probability and estimating its criticality. With this information, the security manager will be better equipped to justify the allocation of resources (budget) to mitigate the effects of these hazards. "By analyzing statistics and methods of operation, specific crime-conducive conditions will be obvious."[1] The risk for crime can come from within the organization or from third parties on the outside. Because business owners owe a duty of care to employees, patrons, and guests, this chapter also outlines methods to establish whether the business or landowner is "on notice" that future crime is foreseeable and therefore required to take action to prevent the crime, to mitigate its effects, or to warn patrons and employees of the danger.

Unfortunately, there is no exact formula the security manager can use to make these determinations. This is due in part to the following:

- Differing laws among states
- Differences in the duty owed and the interpretation and applicability of case law for various types of property. Restaurants, shopping malls, and theaters, where guests are "invited," may be treated differently by the courts than an industrial site, campus, or doctor's office.
- Changes in case law and decisions
- Different interpretations regarding the assignment of responsibility. Some courts believe that if the perpetrator of the crime is at least 50 percent responsible (negligent), third-party liability is dismissed. Other juries have awarded victims large sums although they found the perpetrator only 35 percent responsible.

---

[1]F. J. D'Addario, *Loss Prevention through Crime Analysis* (Butterworth-Heinemann, Boston, MA: National Crime Prevention Institute: 1989).

The prediction of crime is, like risk analysis, an inexact science and is often based on the professional's best educated guess. The methodologies used to predict crime are subject to more debate than those of risk analysis. Many factors influence criminal behavior. Precise prediction is difficult, if not impossible. Statistical models to predict crime, such as the Burgess method, configurational analysis, multiple regression, multidiscriminant analysis, and log-linear analysis, are usually not required to meet the goals of the security manager. Simple methods exist that can give the security manager a systematic indication of future crime risk. Relying on the criminological theory that recidivism is the best predictor of future crime (though some theories hold that age, demographics, or causation are the major predictors), we can apply this rate to individual categories of crime to estimate the potential that crime in the surrounding community will "spill over" and affect the safety of employees at a target location.

ANALYSIS OF INTERNAL CRIME

As in risk analysis, the prediction of internal crime relies on historical data. We can expect past or current crime rates to continue or increase into the future if conditions responsible for the criminal activity (opportunity, for example) do not change. The probability of hazards and events, to a great extent, depends on consistent conditions over a time period sufficient to draw statistical inferences from past data. For example, if a company averages X fires per year, we can predict that the company will probably experience close to the same number in the future if no factors are introduced to mitigate this risk.

Crime analysis focuses managers on the past, not the present. State or federal Uniform Crime Reports (UCRs) can be used as data sources, but their information is already a year old. By the time countermeasures are devised, capital budgets approved, and service or equipment proposals submitted, we implement countermeasures based on data that are 2 years or more out of date. Projecting crime trends will give management a better understanding of present and future crime exposure.

Given that a small percentage of employees cause most crime, the arrest or departure of a single employee could radically affect future rates. Many security professionals believe internal crime, especially theft, is the result of drug- and alcohol-dependent employees. The introduction of pre-employment and random drug testing, education, and treatment programs will surely affect the future projection for drug use, injuries, and theft.

If the business has already compiled adequate historical data, it becomes a simple matter to project the trends into the future. Software programs are available that sort incidents by building, site, and time. They generate incident reports and produce trend analyses by location, day, time of day, modus operandi, and other identifiers to pinpoint expected losses in specific areas or divisions, or by types of crime. If accurate records have not been maintained within the security department, check with the accounting, human resources, internal audit, and risk management departments for information. When gathering data, especially if it is anecdotal, use caution that it is not misreported or misclassified; many nonprofessionals, for instance, confuse burglary with robbery.

Adjustments must be made for factors that can influence the future occurrence of the crimes or incidents being projected. Staffing increases and changes in workforce demographics and employee attitudes (morale, job satisfaction) must be considered.

How will changing rates of domestic violence against women affect the workplace if the workforce is predominantly female? Do labor contracts expire soon?

Projections can be represented as a rate; UCRs list the number of crimes per 100,000 population or as the number of crimes expected per year. The retail department can express its rate as the number of crimes (such as robbery) for the total number of stores. The use of rates will maintain consistency in comparisons if the base (for instance, the number of stores) increases or decreases. Look for cyclical trends, such as rising theft during the holiday season or after new-product releases. Use this information to pinpoint areas of concern and to utilize limited resources in the best way. Concentrate patrols in high-crime or high-potential-crime areas of a campus, building complex, or parking lot. Focus on certain crimes, such as rape, that the projections identify as tending to occur, for example, more often during certain months of the year.

ANALYSIS OF EXTERNAL CRIME

The intent of this section is to guide policy and decision makers to realistic evaluation of the risk associated with criminal behavior from the community that may affect the health and safety of employees and assets. Its assumptions and generalities are not sufficiently focused to allow predicting the future criminal behavior of specific members of that community, or of a specific applicant, employee, contractor, or former employee. The prediction of criminal behavior is an inexact science, open to many debatable issues; errors in prediction are therefore inevitable. The security professional should evaluate the results of these predictive methods in light of the totality of the circumstances, of currently accepted criminological theory, and of the professional's experience. If adjustments seem necessary, look to the costs and personal losses that a false negative may create. Is it in the best interest of the firm to err on the side of extra protective measures against a rape, even though the analysis may indicate the probability is very low (say, one in 25 years), rather than devote more resources to a higher-cost, higher-probability event, such as shoplifting?

External crime can be analyzed for three purposes. One is to determine, as accurately as possible, the true potential that crime in the surrounding community will affect the health, safety, and assets of the organization. The second purpose is to determine whether the business is legally "on notice" that injury to employees, patrons, or guests, by third parties, is foreseeable. The third is to learn how a business should go about gathering data to support or defend a claim of negligent security after an injury has occurred—that is, how does one determine whether the injury was foreseeable?

Many security managers simply determine the crime rates for their locations and benchmark the data with that of a similar organization or city with comparable demographics and population. When using crime rates or survey data from valid research, always compare apples with apples—measure parking-lot crime against parking-lot crime from other areas, and compare it to the norm for the study areas. For example, a survey found that 35 percent of restaurant industry workers admitted to taking company supplies for personal use. Unfortunately, we cannot expect all restaurant employees to follow the same pattern because the study focused only on fast food establishments.

Prediction of the "spillover" of crime from the surrounding community is based on many assumptions. In an industrial setting, we need to assume that the level of security protection is no different from the local standard. It makes little sense to compare burglary statistics for homes or businesses that have a minimal level of protection with those of a location that has the best locks, the highest fences, the brightest glare lighting, closed-circuit television (CCTV), and an armed security force with attack-trained dogs.

The following method is an attempt to estimate the chance of external crime affecting the workplace, absent any security measures beyond the local standard. This method is based on criminological theories and established trends, but it has not been validated in a court of law or by a professional review board. I have found it both a useful and accurate tool to justify manpower and budget decisions based on projected needs.

1. Select the crimes to measure or predict. Include Category I ("Indexed") crimes:

   - Murder
   - Nonnegligent manslaughter
   - Forcible rape
   - Robbery
   - Aggravated assault
   - Burglary
   - Grand theft
   - Motor vehicle theft
   - Arson

   Include other crime exposures that would have a negative impact on the company, especially those that may result in litigation. These may include simple assault and battery, vandalism, trespassing, drug sales, and the like. Be sure to consider any special risks, such as the presence of local extremist groups.

2. Research crime rates for the offenses selected above. Data on crime rates are obtained from local, state, and federal sources. (See Appendix F, How to Establish Notice, for specific sources.) Understand that "there are concerns that criminal justice data collection mechanisms are woefully inadequate and not standardized across the State and Federal systems."[2]

   Crime statistics are inaccurate because of underreporting and differing interpretations of the definitions of crimes. Increases or decreases in the rates are also affected by the level and focus of enforcement (that is, targeted enforcement) and by the efficiency of the enforcement agencies. The Federal Bureau of Investigation (FBI) UCR summarizes crimes under the "hierarchy rule," recording only the most serious crime within an incident. The UCR is the nation's primary source of information about crime and arrest activities of local law enforcement agencies. It is relied on by the general public as an indicator of

---

[2]The Joint Emergency Preparedness Program (Jepp), *Domestic and Sexual Violence Data Collection: A Report to Congress under the Violence against Women Act* (Washington, D.C.: National Institute of Justice, 1996). Although improved, this is still true today.

community safety. A relatively new system, designed to alleviate many of the current problems with the UCR, "moves beyond aggregate statistics and raw counts of crimes and arrests that comprise the summary UCR program, to individual records for each reported crime incident and its associated arrest."[3]

This program, the National Incident-Based Reporting System (NIBRS), eliminates the "hierarchy rule" and looks at detailed offense, offender, victim, property, and arrest data in 22 crime categories and for 46 offenses. Although much improved, implementation of this system has been slow since its inception.

Determine the number of crimes committed for a ½-mile, 1-mile, or 3-mile radius, as appropriate, possibly adjusting for any natural boundaries that may skew the results, if the local jurisdictions maintain data at this level of detail. Crime-tracking software used by many police departments has the ability to list, tally, and plot graphically the various types of crimes around a specific location. If this is not available, use crime district information or, as a last resort, use citywide statistics. Extract the raw numbers from the rate by multiplying the basis by the rate. Record these numbers over time: the prior 2, 3, 5, or 10 years. Record the percentage increase or decrease in crime rates over each year for the target areas. Use this percentage for the calculation in step 5 below.

3. Adjust the data in each category by the ratio of reported to unreported crime. Victimization and criminological studies conclude that 63–67 percent of violent crimes are unreported. Only 16 percent of rapes are reported to the police. By adjusting the raw figures by these percentages, you should arrive at the "true" (at least a more accurate) rate of crime for each category. Compare your adjusted figures with those derived from victimization studies. This may take some research because these numbers are also subject to change.

4. Reduce the numbers by the recidivism rate (rate of reoffenders). There is discussion about the validity of this weighting because it is not clear whether the 20 percent who do not return to prison have been rehabilitated or are simply evading recapture. How this adjustment will miss new offenders is unknown. Other factors that can affect recidivism include overestimation of the number of reoffenders because of political agendas and racism.

Some believe the numbers should be increased or decreased by changes in the area's overall crime. Simply adjusting overall crime rates up or down according to the recent trends does not account for the variations in the individual classes of crimes. Generally speaking, property and violent crimes are committed by younger age groups, and white collar crime is committed by older age groups. Granted, the

---

[3]J. M. Chaiken, PhD., *Implementing the National Incident-Based Reporting System: A Project Status Report* (Washington, D.C.: U.S. Department of Justice, July 1997).

particular trend for each individual crime can be applied to the data, but this does not consider variations in recidivism for individual crimes. Crimes of violence tend to be impulsive and therefore have lower recidivism (most homicides are one-time affairs), whereas the perpetrators of property crimes tend to continue the practice. The best result is obtained by an examination of the areas' demographics and by adjusting the data for sex and age factors.

5. Calculate the expected recurrence of crime for your location in terms of how many will occur in a certain period (such as 5 years) or in terms of full percentages, such as, "We can expect one rape in the next 15 years, assuming no additional mitigation is introduced."

   When analyzing rates, remember that certain crimes can be cyclical. For example, cases of rape increase in the summer months; vandalism and sabotage occur more often before and during labor disputes.

6. If possible, set probable loss figures for the predicted events. If your analysis predicts that the truck transporting finished goods from manufacturing to the warehouse will be hijacked once every 5 years, and the average or maximum shipment is worth $4.2 million, you can calculate the impact of this loss. Remember to subtract insurance reimbursements and to add contractual penalties for nondelivery of product, potential loss of market share, extra costs for remanufacturing, the cost of post-traumatic stress counseling, and other indirect costs. Financial loss can also result from civil litigation, loss of morale, restricted access during crime scene investigation, clean-up costs (blood and glass), recruitment costs for a new security manager or a replacement for the injured worker, and increased worker compensation premiums.

   Consider short-term or crime-specific items, such as workplace violence or domestic violence spillover. The above method is not very useful for understanding the potential for domestic violence spillover because the comparative "population" is different. In this case, the demographics of the workforce, compared with the rate of victimization, should provide a more accurate projection. As with comparisons discussed previously, the actual projections are affected by the number and type of formal and informal controls in place. Until studies are completed that track the amount of domestic violence spillover and establish the effectiveness of various controls, projections are mostly subjective.

   Methodologies that rely exclusively on historical data or other static factors may not best protect the organization against losses from external or internal influences. Past behavior and historical data are not in themselves predictive of future behavior. The analyst must have the ability to project fundamental dynamic relationships into the future. Thus "clinical" criteria developed from the most recent trends, conditions, and the analyst's experience are needed to add practical value to the results. Reliance on purely historical data may not identify a relatively sudden rise in high-technology invasion-style robberies until the problem becomes widespread. The analyst must use great caution that subjective criteria do not add ambiguity or bias the results by unintentionally or subconsciously weighting factors to justify an agenda, and that the added variables are not redundant.

INADEQUATE SECURITY

The occurrence of crime on property controlled by a business may place its owners "on notice" that the recurrence of a similar crime is "foreseeable." If a person is then injured by a third party, the victim may have a cause of action against the business or property owner for inadequate security. Management ignorance often is responsible for liability in injuries caused by third-party crime. The wise business owner will conduct a foreseeability study to estimate the level of risk and from its results determine what degree of security protection is reasonable. Although there are few facts and valid studies to support expert opinion, attorneys will argue whether the presence of additional security personnel, increased lighting, attention to procedures, or other measures could have prevented an attack.

Many business owners and corporate managers are motivated to conduct a foreseeability study by fear of litigation. Absent litigation, a violent act on the property will almost certainly create an unwanted cost to the organization through reduced employee morale and customer confidence. Additional justifications for a foreseeability study are that its results may accomplish the following:

- Lead to a summary judgment of a case
- Pinpoint areas of concentration for guard staff and patrols
- Lead to better utilization of resources and equipment
- Improve overall security and financial planning
- Help to justify security budgets and programs
- Aid in site selection for new facilities
- Lead to better understanding of crime risk

Third-party liability cases are expensive and often difficult to defend. Judgments for security-related negligence often exceed insurance coverage. Firms have found themselves underinsured, despite high premiums, or not insured at all. Policies that cover the loss of customers, reputation, or future business do not exist. For plaintiffs to prevail, they must generally demonstrate the following[4]:

- The business or property owner had a duty to protect.
- The business or property owner breached the duty.
- The breach of the duty (crime) was the legal (proximate) cause of the injury.

Crime analysis completed for security-planning purposes is usually not sufficient for presentation to a court subsequent to a negligence (inadequate security) claim. The necessary scope of the analysis, including the type of data analyzed, will change. Property (as opposed to violent) crime, concentration on the specific cause of the injury, and the

---

[4]Exact legal requirements vary among states.

sources of the data become more important. You must match or exceed the sources of information the opposing parties intend to use in their attempt to establish notice. If the opposition bases its analysis on data that include arrest and incident numbers, you need to do the same. (Arrest information, however, is illegal to obtain in certain states. You may be at a disadvantage if the opposition has access to police contacts, but this information may be subpoenaed or discovered by court order.) This is one of many reasons the security manager or investigator must work closely with legal counsel.

In California, the duty to protect ends at a public area, such as a sidewalk or grass strip, unless it is shown that the business or property owner took control over this public area. There are, however, notable exceptions to this rule. Control could become an issue if employees or patrons must walk across an "uncontrolled" (that is, unowned) property to get to the controlled property. The business or property owner does not always need to own, possess, and control the property in order to be held liable; some courts consider that control alone is sufficient, but others hold that no liability is established in this instance. In one court decision, the actual or "apparent" control over immediately adjacent property and the foreseeability of injury created a duty on the part of the property owner to protect the victim from the danger (or to warn the victim of the danger). A duty may exist if the design of the building or passageways forces employees or patrons to pass through dangerous areas.

As a security manager, use a liberal approach in the analysis. Although the courts may rule in favor of the business in a given case, the time, expense, and adverse publicity of such litigation is to be avoided. The control of an area is usually a triable fact. Additionally, courts view the issue of derived benefit to the injured person differently. At least one court takes the position that "liability does not depend upon whether the defendant derived a commercial benefit from the property" (*Princess Hotels International, Inc., v. Superior Court,* 33 Cal. App. 4th 645 [1995]). It often comes down to what testimony the judge will include or exclude in establishing control and foreseeability for a reasonable distance from the property. For planning purposes, consider crime at adjacent properties even if no legal control over the property exists. Although you may not be responsible in court, an injury could have a damaging effect on your employees.

Once duty and notice are established, it becomes a matter of causation—what level of security protection is reasonable, and would it have deterred, mitigated, or prevented the offense?

There must be a causal connection (proximate cause) between the negligence to protect and the injury or attack. Proximate cause is often difficult to prove—for instance, did the lack of additional security officers or the lack of increased lighting contribute to the attack? This may be difficult to prove or disprove in court. Standards for the amount and type of security the court will find reasonable vary from one type of property to another. Whereas the court may find a shopping mall liable for not providing uniformed security officers, they may rule differently (find no requirement for security officers) in the case of a small business. "Standards or minimums for security can never be ironclad. Security procedures must be adapted to local conditions and changes. That is why the trial attorney requires a security expert to define security negligence."[5] The exact character of the injury is not the correct standard; rather, the question of foreseeability must be decided by the type of harm likely to be sustained.

---

[5]N. R. Bottom, Jr., PhD., *Security Loss Control Negligence* (Maryland: Harrow Press, 1985).

HOW TO ESTABLISH NOTICE

Notice is based on prior similar incidents. A high rate of embezzlement or other type of white collar crime does not place the business or property owner on notice for a rape. This does not mean that a high incidence of robbery does not put the owner on notice for rape. Courts have concluded it is not necessary to decide whether particular criminal conduct establishes notice, but rather that it is necessary "to evaluate more generally whether the category of negligent conduct at issue is sufficiently likely to result in the kind of harm experienced that liability may appropriately be imposed on the negligent party" (*Ballard v. Uribe,* 41 Cal. 3d 564, 573, fn.6 [1986]). Also, "It is possible that some other circumstances such as immediate proximity to a substantially similar business establishment that has experienced violent crime on its premises could provide the requisite degree of foreseeability" (*Ann M. v. Pacific Plaza Shopping Center,* supra. 6 Cal. 4th 679, fn.7, 1993). The dissent to this opinion, shared by other courts, holds that "similar" means "identical" at a very specific location (for instance, a rape next door does not put the establishment on notice for rape, only a rape that actually occurs on the premises). Recent decisions support the majority decision. The California Appellate Court in *Lisa P. v. J. Gordon Bingham* (43 Cal. App. 4th 376 [1996]) held that prior armed robberies were not similar in nature to the rape of a clerk but still had the effect of putting the defendants on notice. Recent court decisions have modified that decision (*Sharon P. v. Arman Ltd.,* 21 Cal. 4th 1181 [1999]), but the manager must be aware of the range of possibilities they may encounter.

A business or landowner who knows, or reasonably should know, that criminal behavior is occurring on or close to his or her premises must investigate to determine whether the criminal behavior is likely to pose a risk to those who enter the property in the future, and whether some aspect or feature on the property (lack of lighting, or inadequate locks) encourages the criminal conduct, or at least makes it easy to perpetrate. This seemingly negates the concept that business and property owners in some jurisdictions receive "one free crime," one for which notice is not established until the crime actually occurs on the controlled property.

Usually the crime that puts the business on notice must be relevant—examine violent crimes in comparison with prior violent crime. However, be prepared to discuss crimes, such as burglary, that, although not violent in themselves, may have a violent outcome.

To establish notice, analyze crime data over varying time periods and distances from the premises. No consistent standard or formula exists to tell the analyst what geographical area and time period to examine. The area and period are best defined by counsel. The analyst will combine the parameters provided by counsel and utilize the various sources of data available to identify the relevant incidents within these boundaries. The analyst will then examine the details of the individual cases to confirm further their relevance to the circumstances that would or would not place the business or property owners on notice. Although anecdotal evidence supplied by employees or others may not be statistically useful, it may be introduced into evidence by the opposing party and would therefore become relevant to the case. The results of the analysis are presented in a report or memo.

After-the-fact analysis of notice is a straightforward process. The challenge arises when the security manager conducts this type of study before any litigation. He or she must anticipate the types of incidents that would cause negligence. The analyst must examine a broader range of scenarios and draw inferences on issues of control, proximate cause, and other negligent security issues. The analyst must consider a wide range of injury, category I crimes, as well as property crimes, for a focused location or for multiple locations, but must also compare these incidents for a more generalized area. A campus with many differing locations may have differing crime rates, security exposures, and conditions that invite crime.

Previous courts have rejected crime in the neighborhood as inadequate evidence for foreseeability inside a shopping mall. It is important to make comparisons as similar as possible—shopping malls with shopping malls, manufacturing with manufacturing. But in these cases, more is better. Include crime in the neighborhood as part of your analysis; it then becomes a matter for the courts to decide what data are relevant. Normally the incidence of crime is examined at the adjacent property or from a 2,000- to 3,000-foot radius. Include a comparison to other cities, counties, states, and regions. If possible, compare individual districts within the city or similarly ranked districts in other cities with the same population.

The number of police calls for assistance in the area during the past 5 years can be used to show notice. Determine the ratio of calls that did or could result in violence to property, such as burglary and vandalism. The opposition will use this to prove, or try to prove, that the business knew, or reasonably should have known, of prior incidents near the business. Courts usually look at a 3-year review of rates of crimes against people (category I). Crimes against people account for 99 percent of the inadequate-security litigation exposure. The courts, however, may consider a 5-year period as reasonable, and plaintiffs have used 10-year periods when it is to their advantage. It then becomes a matter of convincing the judge which study is the most reasonable.

SOURCES OF DATA

Local Police Departments

Check with the crime prevention bureau, department of statistics, administration, or public information officer (PIO). Some departments use software programs that will print color maps of crime incidents for selected distances around a particular address. Check with other agencies, such as transit police, for information on crimes committed in their jurisdictions.

News Media

Newspapers, television, and radio reporters, as well as their archives, are good sources of information. These can be researched in person at the publication's office, through a local library, or by using on-line searching.

Subpoena

This can include police records; crime prevention or physical security surveys completed by police, security, or insurance auditors; insurance loss runs; and electronic mail records.

State and FBI Unified Crime Reports

The FBI compiles crime data from across the nation and reports it by city, region, category, age, and other categories. A copy of *Crime in the United States* can be found at the local library or at the FBI's or other World Wide Web sites.

ATF/FBI Arson and Bomb Reports

Publications from the Bureau of Alcohol, Tobacco, Firearms and Explosives (ATF) and the FBI contain information on the prevalence of arson and bombing across the United States. The National Fire Protection Association in Avon, Massachusetts also maintains data on suspicious fires.

Victimization Studies

Victimization studies rely not on arrest or conviction information but on surveys of the general population's experience with crime. See the U.S. Department of Justice *National Crime Victimization Survey*.

Valid Internal or External Surveys

Survey the employees about their experience with crime in and around the business. Ensure that the results are statistically valid.

Centers for Disease Control and Prevention

The Centers for Disease Control and Prevention (CDC) maintains information on a range of topics, including violence in the workplace, the use of firearms, and other violent crimes.

Local College Campuses

Although most college campuses are required by law to maintain crime data, crime committed on college campuses may not be reflected in local reporting sources.

Canvassing

Anecdotal information from interviews with patrons, employees, and community members is useful as a double check and can lead the investigator to sources others may miss. Speak with the following community members:

- Neighbors
- Competitors
- Fire and ambulance crews
- Union representatives
- Security officers
- Postal carriers
- Regular delivery drivers and suppliers (Federal Express, UPS)

Public Library

Many of the sources listed above can be found at the public library. You can also research past news articles for information at the library or connect through their web sites to clipping services.

Surrounding Businesses or Corporations

Examine their incident reports and records, and interview longtime employees who would have knowledge of crime, such as the human resources director, security manager, or insurance/risk manager for insurance loss reports.

Bureau of Justice Statistics

The NIBRS, *Criminal Victimization in the U.S.*, *Violence and Theft in the Workplace*, and *Sourcebook of Criminal Justice Statistics* are some of the useful databases and publications available from the Bureau of Justice Statistics in Washington, D.C.

INCIDENT CLASSIFICATIONS

The following list can be used to help track the occurrence of crimes and incidents. Additional subheadings or subclassifications for crimes committed by employees (internal) or for crimes committed by customers or guests (external) can be included, depending on the needs of the firm. "Robbery" can be further divided by including theft by pickpocket (not robbery in some states), purse snatching, or other divisions.

- Arson
  Actual
  Suspected
- Assault
  Simple
  Aggravated (weapon)
- Attempted rape/rape
  By employee
  By non-employee
- Sexual battery
- Burglary
  Attempted
  Forced entry
- Computer-related crimes
  Attempted break-in
  Disclosure of passwords
  Denial of service
  Web vandalism
- Embezzlement
  Money laundering
- Kickbacks
- Technology transfer
- Extortion

- Forgery
- Counterfeiting
- Misappropriation of funds
- Domestic violence spillover
- Insurance (worker compensation) fraud
- Homicide
- Kidnapping
  Perpetrator known to victim
  Perpetrator unknown
  Executive or key employee
  Attempted
  Threatened
- Theft
  Auto
  Proprietary information
  From auto
  Funds
  Product
  Diversion
  Misappropriation
  Raw materials
  Precious metals
  Personal items
- Disturbance
- Disorderly conduct
- Sabotage
  Suspected
  Product tampering
- Vandalism, malicious mischief
  Vehicles
  Tagging, graffiti
- Suspicious circumstances
- Indecent exposure
- Possession or disclosure of objectionable material
- Sexual harassment or unwanted advances
- Tailgating
- Corporate rule violations
- Parking violations
- Vehicle violations
- Vehicle towed
- Fire access blocked
- Other vehicle code violations
- Substance abuse
  Possession
  Sales
  Under influence
- Hit and run

- Property damage
- Robbery
  Strong arm
  Weapon
  Force/fear
- Suicide
  Actual
  Attempted
  Threatened
- Access control
  Attempted entry
  Unauthorized entry
  Badge missing/stolen
  Misuse of badge/card
  Loaning access control card
- Bombing
  Explosion
  Incendiary
  Threat
  Intelligence/information
- Trespassing
- Prostitution
- Gambling
- Gang activity
- Alarms
  Security
  Fire
  Environmental
  Process control
- Maintenance
  Lighting
  Fencing
  Locks/doors
  Glazing
  Doors/gates/windows open
  Shrubbery/landscaping
- Escort requests
- Police contacts
- Solicitation/special interest
- Telecommunications fraud
- Misuse of company equipment/services
- Obscene/harassing phone calls
- Demonstrations/picketers
- First aid/medical
- Safety hazards
- Terrorist threat
- Stalking

11 ⬚⬚⬚
⬚⬚⬚
⬚⬚⬚

Determining Insurance Requirements

"Insurance has become part of the overall loss prevention plan supporting proper security ... the more comprehensive the security plan, the lower the cost of the insurance."
—*Robert J. Fisher and Gion Green, Introduction to Security,* Seventh Edition, Butterworth-Heinemann, 2004

Having once been employed by what the *Wall Street Journal* described as "the world's largest insurance brokerage firm," I am aware of just how little I know about the complex business of insurance. Nevertheless, working with some of the industry's most outstanding brokers and risk managers from a number of Fortune 500 companies does give one an appreciation for, and some insight into, the vital role insurance plays in risk analysis and management. The reader is cautioned, however, that this chapter is only a brief introduction to some aspects of insurance of which the security professional should be aware. The advice of competent insurance professionals should always be obtained before deciding on, recommending, or considering insurance matters.

RISK MANAGEMENT DEFINED

Risk management can be defined as the process by which an entity identifies its potential losses and then decides what is the best way to manage these potential losses. Once a risk is identified, analyzed, and evaluated, the optimum method of treating the risk can be chosen and put into effect. Security-related losses can occur as a result of a variety of factors, such as internal and external theft and manmade or natural disasters. Also, losses through fire, safety problems, and product or third-party liability are some of the leading concerns of corporate risk managers.

A properly performed risk analysis can be used for many things, but its end result is a definition of the effect risks have on a particular company, in terms of the potential for loss. The analysis should tell where, when, and how the risk is likely to be incurred. It should also indicate the extent of loss or liability if the risk does in fact occur and how badly the company would be injured. The risk manager can then design a program to cover the potential losses, exposures, and liabilities. In dealing with most risks, the company is faced with three basic options:

- The risk can be avoided, eliminated, or reduced to manageable proportions.
- The risk can be assumed or retained.
- The risk can be transferred to a third party. (Transfer to a third party generally implies transfer of liability to an insurance carrier.)

RISK CONTROL

The process of *eliminating* or reducing risk to manageable proportions is somewhat self-explanatory. This is usually done by programming security and safety procedures to do away with problems or to reduce them to acceptable or manageable levels of severity. This is also referred to as *loss prevention or control.*

By *assuming* the risk, the company makes itself liable for the loss, if any, to be incurred. If the potential loss is deemed to be within the limits of an expected and otherwise acceptable loss, the risk may be acknowledged and left alone. No effort is made to control, eliminate, or minimize the risk. No action is taken to correct the situation, and no insurance is purchased to cover it. In some cases, a company may develop some form of self-insurance, whereby the exposure or liability is assumed by the company itself. In most instances of risk assumption or retention, the risk is perceived to be small enough that management is willing to assume total responsibility and absorb, out of operating expenses, any losses that may occur—the rationale being that the cure would be worse than the disease. This is especially true in retail stores in which the costs of some losses can be passed on to the consumer.

When a company *transfers* a risk, the risk manager, who usually works in conjunction with an insurance broker, endeavors to find the best insurance program available from carriers in the marketplace that provide the needed type of coverage. This is no simple task; it includes, among other things, determining the best deductible and premium payments available, which in the case of large companies often run into the millions of dollars annually.

Risk management is constantly faced with the problem of the selection of the best method (or if necessary, some combination of methods) of handling each identifiable risk. Regardless of the method or combination of methods used, some basic considerations affect most insurance programs. To be insurable, risks must substantially meet the following requirements:

- The risk should be worth the cost and effort to insure.
- The risks are calculable, through large numbers of similar risks.
- Losses can be clearly established as to occurrences and amounts. (This is especially important to the security professional who usually investigates the loss in question.)
- Losses must be accidental in nature, unexpected, and unintentional on the part of the insured.

CRIME INSURANCE

Crime insurance is usually obtained to supplement a company's security program. Although the presence or absence of insurance has no deterrent effect on crime, it does reimburse the company in whole or in part for losses sustained in a burglary or robbery or from internal theft. Crime insurance should, like other coverage, be tailored to meet the specific needs of the client. As one insurance broker advises, "If you want to know what your crime insurance covers, ask what it doesn't cover."

At a minimum, most crime insurance programs begin with a "3D policy"—comprehensive dishonesty, disappearance, and destruction—a blanket crime or broad-form storekeepers' policy. This coverage will usually reimburse a company for losses due to employee dishonesty or counterfeit currency, as well as for loss of money, securities, or merchandise through robbery, burglary, or mysterious disappearance. These policies also generally cover certain types of check forgery and damage to the premises or equipment resulting from a break-in.

Some forms of specialized crime insurance coverage usually available for consideration include the following:

- *Mercantile safe-burglary policy:* covers loss of money, securities, and valuables from a safe or vault, and pays for damage to the container and any other property damage as a result of the burglary.
- *Mercantile open-stock policy:* mostly used by retail firms as coverage against burglary or theft of merchandise, furniture, fixtures, and equipment on premises, pays for damage to property resulting from burglary.
- *Fidelity bonds:* reimburse the employer for loss due to embezzlement and employee theft of money, securities, and other property. (Bonds cover certain positions; employees who handle money, cash receipts, and merchandise are usually bonded.)
- *Forgery bonds:* reimburse merchants and banks for any loss sustained from the forgery of business checks.

Insurance premiums vary for these programs according to the type of business, store location, number of employees, maximum cash values, amount of security equipment (such as alarms) installed on premises, and prior losses. Merchants operating in some high-risk crime areas and thus needing insurance are often the least able to afford the premiums. Further, it is difficult to find insurance companies willing to underwrite crime coverage in high-risk crime areas. Companies that experience a number of robberies or burglaries usually face escalating premiums or, worse, canceled policies.

For many small businesses and commercial enterprises, insurance and an antiquated burglar alarm may be the only form of protection affordable. For large companies and corporations, things are much different.

It is generally believed by knowledgeable corporate management, especially risk managers, that insurance should be used for protection only against risk that cannot be avoided or controlled through the effective use of property, casualty, and security techniques. (In the insurance industry, property protection is synonymous with fire, and casualty is synonymous with safety.) This change in philosophy has come into vogue because more and more managers are getting the message that loss prevention, through risk avoidance, elimination, or control, is the best approach to the preservation of corporate assets. It is also recognized that most insurance programs do not fully compensate a firm for a loss, regardless of the coverage.

A vice president of a large California construction company complained bitterly because the police were unable to send a detective to one of the firm's construction sites—a newly developed industrial park—to take a report and conduct an investigation into the theft of a $15,000 compressor that had been delivered the day before. The police officer explained that he would have to take the report over the telephone. Further conversation with the officer revealed that little active investigation would be conducted in a case of this nature: it was regarded, the officer said, as "a problem between you and your insurance company." What the officer was not aware of, and what the vice president soon found out, was that the construction company's insurance policy had a $100,000 deductible clause—the theft of the $15,000 compressor would not be covered. The big problem, the vice president lamented, was not so much the cash outlay for a new compressor as the time it would take for delivery, which was going to cause him major difficulties in meeting his construction deadline.

What this example illustrates is that management must become more interested in avoiding loss, and not rest with the comfortable thought that the company is insured against "any and all eventualities." There is no insurance company I am aware of that would insure this vice president against the mental aggravation he went through; he is now, however, a believer in loss-control procedures as a solution to these kinds of problems.

A word about deductibles: these clauses are intended to reduce the cost of insurance premiums by deliberately excluding small, frequent losses while covering large, serious ones. The premiums are less expensive for two reasons: small claims are excluded if they fall under the dollar amount of the deductible, and the carrier's administrative cost of settling claims is also reduced.

It is neither the intent of this text nor the purpose of this chapter to do more than explain some basic insurance considerations that most security professionals need to understand in order to do their job properly. Further information can be obtained by contacting the local chapter of the Risk Insurance Management Society (RIMS). The address and telephone number of the RIMS national headquarters is listed later in this chapter. Another good source is the business section of the public library. There is, however, one highly specialized form of insurance coverage that, because of historical developments during the past 40 years and the increase in international terrorism, should be given special consideration in this text—kidnap, ransom, and extortion insurance, known in the industry as "K&R coverage."

K & R (KIDNAP AND RANSOM) COVERAGE

K & R insurance has been offered by Lloyds of London Underwriters for more than 60 years. During the 1970s, as the demand for K & R coverage increased, a number of other insurers entered this market. Generally, there is little to be concerned about with respect to the financial security offered by such companies as the American International Group (AIG), Insurance Company of North America, the Chubb Insurance Company, and Lloyds Underwriters. Premium costs and the scope of coverage afforded under each of the companies' policies are generally the basis for deciding from which carrier to purchase insurance coverage.

Such an analysis is normally done at the time the risk manager or broker chooses to explore insuring this risk. Because there is competition in the insurance business—not only from a premium-cost standpoint but also with regard to breadth of form—any coverage comparison here would serve little purpose. Generally, the basic coverage provides reimbursement for loss of monies surrendered as a ransom payment for actual or alleged kidnapping, or following receipt of a threat to injure or kidnap an insured person. In addition, some unique features, such as the following, are usually incorporated into the policy contract:

- A *business premises extension* reimburses for any monies that must be brought in from outside for any kidnap situation if the money is lost while it is on the premises.
- A *transit extension* reimburses for any monies that are stolen between leaving the premises and reaching the kidnappers.
- A *reward extension* includes coverage for monies paid to informants whose information leads to the arrest and conviction of the individuals responsible for the kidnapping.
- A *personal assets extension* reimburses the insured for their personal assets that are used as a ransom payment if the demand is made on the insured person and not the corporation.
- *Negotiations, fees, and expenses* reimburse for reasonable fees and expenses incurred to secure the release of a hostage, including interest on a bank loan to pay a ransom payment.
- A *property damage coverage extension* provides coverage against threats that cause physical damage to property.
- *Defense costs, fees, and judgments* cover costs resulting from any suit for damages brought by an insured person. The importance of this extension is underlined by an occurrence wherein a kidnapped executive later sued his employer for $185 million in damages, claiming that the employer had not exerted sufficient efforts to free him and had not taken steps to protect him from such an occurrence after being warned that the executive might be a target of abduction.

There are some general requirements, relating to secrecy concerning the fact that the company has K & R insurance coverage, of which the security professional should be aware. Here I caution that the specific details of each policy must be studied and adhered to in the event of a kidnap; otherwise, the incident may be noninsurable. Some of these considerations include the following:

- The ransom or extortion demand must be specifically made against the named insured.
- The extortionate demand must be made during the time frame of coverage as set forth in the policy.

- The company has taken every reasonable precaution to ensure that the existence of the coverage is not disclosed to anyone except senior officials of the corporation.
- If a kidnap occurs, every reasonable effort is made to determine:
  1. That an insured person has been abducted (note: not all policies cover all employees)
  2. That the police or Federal Bureau of Investigation (FBI) has been notified *before* payment and that instructions and recommendations of the police and FBI in the best interest of the victim are accomplished to the extent possible
  3. That the insurance company is notified at the earliest practical time
  4. That the serial numbers of the ransom payment are recorded

Some underwriters (insurance carriers) require that immediately upon obtaining coverage, written policy and procedural guidelines be established to eliminate the possibility of confusion with regard to the handling of these matters. Box 11.1 lists a number of topics that must be considered in establishing procedural guidelines. As will be discussed more thoroughly in Chapter 17, Crisis Management Planning for Kidnap, Ransom, and Extortion, a more prudent course of action is to develop a crisis management program specifically tailored to the requirements based on the insured corporation's requirements. One of the benefits of such planning is to eliminate confusion before the incident occurs. Once a kidnapping occurs, confusion usually reigns supreme if there is no well-thought-out and rehearsed plan. Even under the best of circumstances, there is high drama and many emotional issues involved when a kidnapping is first reported.

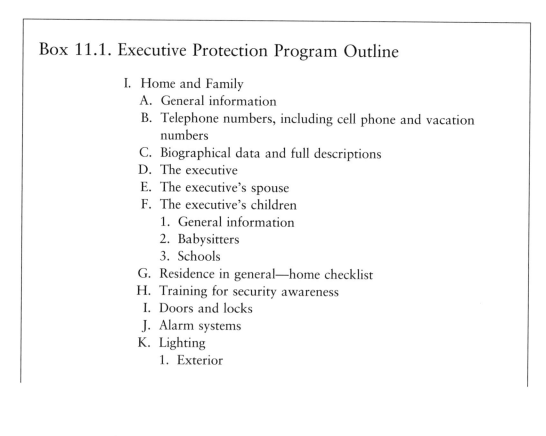

Box 11.1. Executive Protection Program Outline

    I. Home and Family
       A. General information
       B. Telephone numbers, including cell phone and vacation numbers
       C. Biographical data and full descriptions
       D. The executive
       E. The executive's spouse
       F. The executive's children
          1. General information
          2. Babysitters
          3. Schools
       G. Residence in general—home checklist
       H. Training for security awareness
       I. Doors and locks
       J. Alarm systems
       K. Lighting
          1. Exterior

2. Interior
L. Fence and barriers
M. Window grilles
N. Dogs
O. Safe room
II. Office and Work
A. Premises—general
B. Access control
C. The executive
D. Executive profile
E. Employees and associates
1. Office rules and work procedures
2. Meetings
F. Security guards
G. Bombs
1. Surveys
2. Target hardening
3. The bomb incident
4. Letter and package bombs
5. Antibomb curtaining
H. Threats
1. Telephone
2. Written
I. Hostage
J. Hostage calls
III. Travel
A. Automobile
B. Chauffeurs
C. Defensive driving
D. Walking
1. Jogging, golfing, and tennis
E. Elevators
F. Taxicabs
G. Aircraft
1. Company
2. Commercial
H. Overseas or long distance travel
I. Reservations
IV. Personal Protection
A. Firearms—defense
1. Laws
B. Choosing a defense weapon
C. Firearms proficiency
D. Weapon—method of carrying
E. Bodyguard

(Continued on next page.)

Box 11.1 *(Cont.)*

          1. Selection
          2. Personal qualifications
          3. Professional qualifications
          4. Guarding—locally
          5. Guarding—away from home
          6. Visitor protection
          7. Motorcades
          8. Public appearances
    F. Protective clothing

V. Crisis Management Team (CMT)
    A. Defined
        1. Purpose
        2. Composition
        3. Scope
    B. Organization and planning
        1. Readiness plan (prevention)
        2. Contingency plan
           a. Worst possible case scenario
        3. Training the team
    C. Intelligence training
    D. Law enforcement liaison
    E. Public relations considerations
    F. Ransom
        1. Policy and procedures
        2. Limitations
        3. Negotiations
    G. Civil liability considerations
        1. Injury of employee(s)
        2. Wrongful death claim
        3. Stockholder suits

These are some subjects that may need attention. The list is by no means exhaustive. It is our experience that a comprehensive K & R plan should be tailored to the organization, its executives, and the personnel insured. We have found no other way to ensure that all the bases are covered. And, once the plan is developed, it must be constantly reviewed and updated as events and personnel change. An outdated plan is worse than no plan at all.

For serious students of risk management and security professionals who want to understand better how insurance relates to their jobs, we recommend contacting Risk Insurance Management Society (RIMS) Publishing, Inc., 1065 Avenue of the Americas, 13th Floor, New York City, NY 10018 at (212) 286-0202 and subscribing to *Risk Management,* a monthly magazine published by the RIMS.

Emergency Management and Business Continuity Planning

12

Mitigation and Preparedness

Eugene Tucker, CPP, CFE, CBCP
Contributing author

"When I said my business was preventing disasters, a woman from Hong Kong asked me, 'How can you stop a typhoon?' Of course, I can't, and neither can anyone else on this planet."
—John Laye, FBCI Contigency Management Consultants, Morage, CA

Mitigation and preparedness are two of the four components that compose the concept of *comprehensive emergency management* (CEM). The remaining components are response and recovery. Because the term has its origin in governmental emergency management, *recovery* in this sense means the reestablishment of infrastructure, the local postdisaster economy, and other associated issues; however, it can also include the terms *disaster recovery* and *business continuity*.

CEM is an integrated approach to the management of emergency programs and activities for all emergency phases, for all types of emergencies and disasters (natural, manmade, and attack), and for all levels of government and the private sector. It provides a framework for a complete planning process that avoids the tendency to plan for only one element. Very often, a plan or organizational responsibility will focus on a single element such as business continuity (recovery) and not incorporate the necessary elements of mitigation or response, for example. When one person is responsible for emergency response and another for business continuity, the ineffective use of resources, overlaps in effort, and disconnects in the transition from one phase to another are typically found. A truly effective plan will include each component. It adds flexibility to the process by allowing the organization to deal with an incident of any size or complexity.

The concept of CEM is also a risk management method increasingly applied to security planning. The various ways to treat risk—risk avoidance, risk assumption, risk transfer, and risk control—are in one way or another addressed by the CEM model. It forces managers to think in a less myopic way when devising a complete risk-based loss prevention and protection of assets or counterterrorism program.

MITIGATION

Mitigation is sustained action that reduces or eliminates long-term risk to people and property from natural hazards and their effects. According to the Federal Emergency Management Agency (FEMA), *mitigation* refers to specific actions that can be taken to reduce loss of life and property from manmade hazards by modifying the built environment to reduce the risk and potential consequences of these hazards. It is vulnerability reduction; it reduces the potential for future losses. It is a strategy to eliminate or to reduce the impact of any impediment toward reaching a goal. Its definition and usefulness go beyond natural hazards to include manmade hazards—a definition that is now expanded to include technological hazards and terrorism, such as the effects of hazardous materials accidents and the use of weapons of mass destruction. The concepts of crime prevention and prevention in general fall within this category of CEM.

Mitigation is often a concept underutilized by the business continuity planner, but its importance in many aspects far out-shadows its focus on simply producing a continuity plan. A good mitigation program that eliminates or reduces the risks can prevent the need to implement a continuity plan in its entirety. The investment in a good mitigation program can return literally millions of dollars in damages avoided and help to ensure the survival of the firm after a disaster. It makes communities and businesses disaster resistant and can reduce damage to property and other assets. A good example described in the FEMA literature[1] is the case of the Kingsford Manufacturing Company's charcoal plant, which sustained $11 million in flood damage and did not return to full production for 6 months after a 2-month shut down. Eleven years later, the plant suffered another $4 million in damage, again the result of flooding (it was shut down twice during that year). The company invested $2.85 million in mitigation (the construction of a levee and other measures) and has avoided not only additional major flood damage but also relocation, helping to sustain the economy of the area. Because the company was able to remain in production in subsequent years, it avoided the loss of retail shelf space (the desirable location and amount of space in a supermarket shelf that draws the most attention to the product), which can take years to regain.

Mitigation can reduce the occurrence of a hazard or a loss. Stronger building codes may not prevent an earthquake but can drastically reduce the damage caused by one. Mitigation can reduce exposure to civil or criminal liability in the event of a terrorist attack or technological accident. Mitigation actions may help reduce insurance premiums. Mitigation allows for a smoother and therefore faster recovery, reduces the "oops" factor as unforeseen incidents and consequences are reduced, and as previously mentioned, helps to avoid the need to restore operations.

Mitigation is cost-effective. According to FEMA, mitigation measures increase construction costs for new facilities between 1 and 5 percent, but when considering earthquake remediation, the cost can be five times higher than the original cost. For every dollar spent on mitigation, three are saved on damages avoided. Warner Brothers estimates they saved $1 million in losses during the Northridge, California earthquake

[1]Protecting Business Operations, FEMA Publication 331, August 1998. (Washington, D.C.: US Government Printing Office).

because of their mitigation efforts. A company that manufactures equipment for the paper and feed industry spent $40,000 in flood mitigation and saved $3 million after Hurricane Eloise, which included $2 million in lost revenue avoidance. Seafirst Bank spends $17.00 to prevent the loss of $3,000 computer systems, estimating a 4 to 5 percent mitigation cost versus replacement, saving $30 million in replacement costs.

The cost/benefit ratio of mitigation strategies is more difficult to determine when dealing with terrorist acts because, absent very specific and credible intelligence, the recurrence rate is more of a guess. Quantification of the damage may not be known, unless you can use maximum probable loss or engineering modeling, and the time period of exposure may vary. Deterrence has always been difficult to quantify.

Mitigation is generally considered the first phase of CEM, but it is not a linear process. Many believe it is important to integrate the recovery and mitigation phases, but acknowledge that mitigation takes place during the other three phases of emergency management. They point out that after a disaster, the availability of funds and interest in taking action is at its highest. The need to make repairs also signals a good time to build-in mitigation measures, the cost of which may be partially absorbed by assistance programs. After the effects of the disaster are stabilized, one can analyze the damage and design strategies to prevent a repeat of the disaster's consequences. In a business environment (and in the general community for that matter), there is sufficient history, knowledge, and experience to identify beforehand the types of hazards that may affect your location and to then predict their likely effects. You don't need to first break your hand before you learn to wear boxing gloves in the ring. Similarly, you don't need to watch your facility burn down before you can consider installing an automatic fire sprinkler system. Mitigation should start the process and end the process. If, after you identify what to mitigate and then go through recovery and find that something was missed, you go back to mitigation. After action analysis can be considered a form of mitigation if you identify a better method or hazards that were not anticipated. Mitigation planning is often a part of the business impact analysis phase of the business continuity planning process (see Chapter 14), especially if using an outside consultant, because the methods used to identify and quantify critical functions, the people who should be involved in the process, and the need to conduct inspections are similar and usually the most expedient. The mitigation plan is often combined with the business impact analysis report, which outlines the cost of recovery strategies that are often a form of mitigation.

The FEMA mitigation methodology and much of the FEMA literature are geared toward mitigation planning on a regional basis, not for individual business enterprises. This approach encompasses five major steps: organize resources, assess risks, develop a mitigation plan, implement the plan, and monitor progress. Although the emphasis is on community and governmental protocols, such as memorandums of understanding, and includes steps to satisfy the requirements of the Disaster Mitigation Act of 2000 (in which the U.S. government will fund state and local mitigation projects), many of the methods can be used to develop information that is beneficial to corporate mitigation efforts. For example, the "assess risks" phase includes the following steps:

- Identify all hazards that may affect the community.
- Narrow the list to hazards most likely to cause an impact.

- Develop a hazard profile; that is, determine how bad it can get. This information is used to determine the assets in the hazard area you need to inventory. Areas that can be affected by the hazard are mapped. This tells where the impacts are likely to occur.
- Inventory assets. List assets that will be affected by the event. Governmental planning considers assets to include hospitals, schools, infrastructure and utilities, and the like. Assets affected may vary by the different types of hazards and are inventoried for each. The value of the assets is estimated in this step.
- Estimate the losses. This answers the question, How will assets be affected by the different hazards? The loss of structures, contents, use, and function is determined and summed to arrive at a total loss for each hazard event based on the total or percentage of damage.

Similarly, in a corporate planning process, we identify the hazards; devise strategies to reduce or eliminate the impact of the hazards; select the most practical, cost-effective solution; and gain approval and funding to implement the solution.

Hazard Identification

If a hazard is not identified, it cannot be prevented or mitigated. Many methods exist to assist in the identification of hazards, but few give one a precise formula that applies to all environments or that will help to reveal conditions that are not previously known. This is due in part to the perceptions, values, knowledge, and methodology used by the planner. Unless your processes are very complex, hazard identification in a corporate environment looks outside the organization as well as internally at hazards and risks to processes, materials, and equipment. As discussed later, interviews with the process owners should reveal hazards that may require mitigation. The ability to identify hazards is really a state of mind, based on a healthy degree of paranoia, knowledge of history, and information about cause and effect.

History

It is important to understand the hazards in your community. Even if you are located on high ground not subject to flooding, is the city's sewage treatment plant located in a flood zone and subject to damage in the next flood? Although you may have no flood damage, you may be shut down for health reasons. An examination of the city's past disasters will reveal this concern. Most natural hazards are easy to identify because they often recur on a roughly periodic basis or typically occur in the region under study. If your manufacturing site is located at the base of a mountain that has produced landslides in the past, you may be affected by land movement in the future. Earthquakes recur in the same region according to very broad, often predictable time windows. Major tornados and storms are more common in certain locations. Look at the history of the region and of the specific location and list the type of natural hazards and the magnitude of their effects. Determine the maximum impact of these historical events. Predict the future impact of past events, taking into consideration recent mitigation efforts or conditions that contribute to a greater impact, such as recent construction near coastal waters prone to tsunamis. Research newspapers and other historical records. Libraries are a good source; locally or federally declared disaster reports are another. Review existing plans and reports that can be found on the Internet, land use plans, and geological reports. Interview experts in the community, including Office of Emergency Services (OES) directors, universities, architects, and fire and police officers.

Speak with the "old timers" from the area, or seek the judgment of experts, such as a geologist, to offer their opinion of the stability of a hillside, for example. In the United States, FEMA, the U.S. Geological Survey (USGS), the National Oceanic and Atmospheric Administration (NOOA), the U.S. Department of Justice, and Homeland Security maintain web sites that contain information useful to hazard identification. Insurance loss control histories, experiences of other businesses in the area or in the same industry can answer many of these questions.

Inspections

A basic component of hazard identification begins with a complete physical inspection of your facilities and their surroundings. Look at your facility from both a macro and a micro view. What hazards exist in the community that will affect your site? Is it located in a high crime area? Can nearby businesses or their processes negatively affect your site? If your plant is located next to a munitions or fireworks factory, or if the tax revenue office is located on the next floor, you may find this of some concern. Look at conditions, equipment, and items and visualize ways they may negatively affect your concern (see the section on Cause and Effect, later). If there are trees on the site, do they pose a fire hazard? Can they blow over and injure someone or damage power lines? Can someone climb up to a less protected second story window? Look for items that could cause collateral damage. For example, if someone placed an explosive device next to an ammonia tank, would the resulting release of gas cause a greater problem? A physical inspection should identify nonstructural hazards, such as file cabinets not bolted to the floor or wall in seismically active areas. Is the building's air vent unprotected and accessible to a terrorist? Failure analysis, knowledge of history, and understanding of cause and effect are important skills needed to identify hazards and risks during inspections. The experience of the inspector cannot be understated.

Checklists

Checklists can help with the inspection process by giving the inspector clues about what to look for and can help ensure that all questions and areas of inspection are covered. Checklists, however, are not the only tool used for a complete identification of hazards. Checklists can never be complete because the number of environments and hazards are varied and numerous. They should be used as reminders and as a springboard for further thought and should include items that answer the following questions:

- How can employees be injured?
- How can critical systems be damaged?
- What single points of failure exist?
- What hazards can disrupt operations?
- How will hazards effect the environment?
- What hazards can have a public relations impact?
- What hazards might generate regulatory issues?

The checklist should view perils under both normal and disaster conditions.

HAZUS

HAZUS (Hazards United States) is one of many risk assessment software programs that analyzes potential losses from earthquakes, floods, and hurricanes. It is used with Geographical Information System (GIS) software to map the effects of disasters and can estimate damage from these hazards before, during, and after the event. Depending on the skill of the user, HAZUS can estimate physical damage to residential and commercial buildings, infrastructure, and other critical facilities; calculate the economic loss from business interruptions, repair and reconstruction costs, and lost jobs; and determine the social impacts of the disaster. It can be used to model technological hazards (nuclear and conventional blast, radiological, chemical, and biological) that supplement the natural hazard loss estimation capability. HAZUS is available free of charge from FEMA.

Process Analysis

To excel at this important task, one does not generally need to use each formal methodology, such as Hazard and Operability (HAZOP), Failure Mode and Effects Analysis (FMEA), Preliminary Hazard Analysis (PrHA), or others used in the engineering and safety fields, but if your processes are very complex, you may want to seek outside assistance.

The concept of a HAZOP study involves investigating how process might deviate from the design intent. It uses a multidisciplinary team methodically that brainstorms the design, following a structure provided by guide words and the team leader's experience. *Guide words* are simple words used to qualify or to quantify the design criteria and to stimulate the team's thinking toward the discovery of deviations. Guide words include "no," "more," "as well as," and "other than." They are used to ensure that the design is explored in every conceivable way. The causes of meaningful deviations are identified and their significant consequences mapped.

FMEA is a structured technique used to analyze a design or a process to determine shortcomings and opportunities for improvement. It is a tool to identify the relative risks that are designed into a product or process and to initiate actions that reduce the risks with the highest potential. Risks are rated relative to each other using a Risk Priority Number (RPN) for each failure mode and its resulting effects. The RPN is calculated by the product of the severity rating, the occurrence rating, and the detection rating. A 10-step process is typically used to arrive at a mitigation (control plan) and a recalculation of the RPN to determine its effectiveness.

PrHA is a line-item inventory system of hazards and their risks. It is an approach that identifies known hazards and their potential consequences to develop an expected loss rate (probability stated as a loss event or unit of time multiplied by the potential loss). As with the above methods, it results in a hazard mitigation plan.

Experts

Consultants can provide a fresh outsider's view and identify potential hazards that you may miss. Terrorism, information, or telecommunications experts can make up for any technological shortcomings of the hazard identification effort. Security and business continuity planning consultants and engineering firms abound, with varying degrees of ability, but your insurance carrier's loss control department may have the ability to provide assistance at a reduced cost. Engineering firms can evaluate the structural performance of your facilities during a natural event. Equipment manufacturers may provide ideas or engineering solutions for hardening their products or may provide ideas that other customers have implemented. In some industries in which it may take up to 6 months to replace a piece of equipment or to re-create its environment, and in situations in which the firm is using equipment that is not replaceable, mitigation may be the only continuity strategy available. Department heads or process owners are probably the first source of effective mitigation ideas.

Cause and Effect

A primary goal of security, business continuity planning, and hazard identification is to anticipate the unexpected. Take what you have learned about history, methodology, and your professional experience and use this information to project possible future scenarios. Use scenario planning. "Scenarios form a method for articulating the different pathways that might exist for your tomorrow, and finding your appropriate movements down each of those possible paths. Using scenarios is rehearsing the future. You run through the simulated events as if you were already living them. You train yourself to recognize which drama is unfolding. That helps you avoid unpleasant surprises, and know how to act."[2] Scenario planning is a discipline that all security managers and continuity planners should master. It is a tool to identify and devise strategies based on future variables or changing conditions. Use it to uncover the sources of a crisis, whether a physical or technological hazard. If you are responsible for a petroleum refinery and you are faced with a fire and explosion, you may also need to deal with a toxic vapor cloud release, injuries and damage in the surrounding community from the cloud, injuries and medical response in the refinery caused by the explosion that could overtax emergency services in the community affected by the incident, hazardous material cleanup, control of contaminated water used to put out the fire, and so on. Scenario planning will help you identify all of these concerns. It may be frustrating, however, that management may not agree to fund mitigation projects based simply on your good imagination.

Unlike natural hazards that follow the laws of nature, occur more often, and are therefore more predictable, the acts of a terrorist are more difficult to anticipate. Because targets are often mobile and the terrorist can select those most vulnerable and that return the highest "yield" to their objective, many argue that the threats they pose cannot be identified or predicted. Others argue that it is possible, at least to the extent that we can minimize the damage. Again, scenario planning is useful to make these predictions.

Terrorist acts that are easier to predict deal with those who think rationally, as opposed to those whose thought processes are on the fringe or delusional. Organized groups such as Al Qaeda, the Irish Republican Army, and others tend to follow the same methods and use the same destructive tactics throughout their existence. Kidnapping may be the preferred method of one, whereas bombings may be used by another group, often with little variation. As with any type of crime, we can try to put ourselves in the place of the criminal or terrorist and devise not only targets and methods of attack but also ways to prevent their occurrence or to mitigate their effects. To do this, especially when dealing with terrorist, hate, antigovernment, or apocalyptic groups, we must place their frame of reference, that is, their way of thinking, into ours. Their culture, belief systems, and

---

[2]P. Schwartz, *The Art of the Long View* (New York: Doubleday Dell Publishing Group, 1991).

values could be far different from ours. In devising these scenarios, we must also determine what is and is not technically possible. Keep in mind that Aum Shinriko, a Japanese apocalyptic group, employed top scientists and spent $6 million in their failed attempt to kill thousands in the Tokyo subway (12 people died). Much information is available on the technical difficulties of the delivery of weapons of mass destruction. Beware, however, that the terrorist can probably figure out ways around technical difficulties, or needs to get lucky only once.

Methodology

Humans are creatures of habit, and many organized terrorist groups or criminal syndicates tend to operate in the same manner throughout their history. Certain animal rights factions tend to release animals; others use arson, and still others use explosives to make their point. The Abu Sayyaf Group in the Philippines specialized in kidnapping and murder for ransom before their involvement with Al Qaeda when they modified their tactics to include bombings. When identifying hazards and risks from criminal groups and terrorists, understanding their methodologies will direct you toward mitigation specific to their attacks.

Homeland Security

When identifying the hazards posed by terrorism, the U.S. Department of Homeland Security's[3] methodology looks at four conditions:

- Application mode (the event or hazard)
- Duration (the length of time the target is affected by the application mode)
- Dynamic/static characteristics (the tendency of the event or hazard to change in relation to time, magnitude, or area at risk)
- Mitigating and exacerbating conditions (conditions that can reduce or increase the effects of the hazard). Examples of mitigating conditions include earthen berms that can provide protection from bomb blast effects; exposure to sunlight that can render some biological agents ineffective; and effective perimeter lighting that can prevent or minimize the likelihood of someone approaching a target unseen. Examples of exacerbating conditions include depressions or low areas in terrain that can trap heavy vapors, and a large number of objects in front of a building or facility, such as trash receptacles, newspaper vending machines, and mail boxes, that can provide hiding places for explosive devices.

When identifying threats and vulnerabilities on a macro scale (i.e., as a governmental entity or infrastructure concern), address the following issues of *inherent vulnerability* (the threat that exists independent of any mitigation or protective measures due to the nature of the built environment, such as a target-rich subway station) and *tactical*

---

[3]U.S. Department of Justice, *Vulnerability Assessment of Federal Facilities* (Washington, D.C.: US Government Printing Office, 1995).

vulnerability (the threat that exists due to the presence or absence of mitigation or protective measures that make a target more or less vulnerable):

- *Visibility:* How aware is the public (and therefore a terrorist) of the existence of the facility, site, system, or location?
- *Utility:* How valuable might the place be in meeting the objectives of a potential terrorist or saboteur?
- *Accessibility:* How accessible is the place to the public?
- *Asset mobility:* Is the asset's location fixed or mobile? If mobile, how often is it moved, relocated, or repositioned?
- *Presence of hazardous materials:* Are flammable, explosive, biological, chemical, or radiological materials present on site?
- *Potential for collateral damage:* What are the potential consequences for the surrounding area if the asset is attacked or damaged?
- *Occupancy:* What is the potential for mass casualties based on the maximum number of individuals on site at a given time?

These threats are then ranked according to the levels of the above categories to determine how critical the threats are to themselves and to the public. This allows the planner to set mitigation priorities that are appropriate to the greatest and most credible vulnerabilities of the most critical assets. Be aware that mitigating one target against terrorist activities may expose another, less protected target to additional risk.

Mitigation Strategies

To help find solutions, investigate what others, especially those in your industry, have done, if anything, to solve similar problems, risks, and hazards. The Internet is a rich source (www.fema.gov) of information. Federal, state, and local emergency services organizations often have information and publications that describe specific mitigation strategies.

Mitigation strategies are both general in nature and specific to the hazard. Some specific mitigation considerations are included in Chapter 13, Response Planning listed under the hazards discussed. Use these as a guide to developing your mitigation program and as a foundation to expand on the ones already presented.

General mitigation strategies can be loosely classified under the following headings. These categories are presented only to give the reader an idea of types of strategies available:

- Risk management
- Engineering controls
- Regulatory controls
- Administrative controls
- Service agreements
- Redundancies and divergence
- Separation of processes

Risk Management

The principals of risk management can be used to identify effective mitigation strategies. The *hierarchy of control* holds that the elimination of a hazard (risk avoidance) is the first and most effective method to control a hazard. If the hazard no longer exists, you don't need to worry about it. Relocating your facility on higher ground further from a river and moving operations to a lower risk area are examples. The hierarchy lists the preferred order of controls, from the most effective to the least effective; elimination, substitution, engineering, and administrative controls are generally the steps followed in the hierarchy.

Substitution involves replacing a hazard with a process that is less hazardous or nonhazardous. A high-technology manufacturer used a chemical that was making its workers sick. The company found a way to produce the product with a less toxic substance and saved many dollars in health costs.

Engineering Controls

Absent the complete removal of a hazard (which is usually impossible or impractical), engineering controls are probably the best form of mitigation, especially when compared with codes, standards, and administrative controls, because they do not generally rely on human intervention or action. The flood diversion dam discussed earlier is a good example. Bomb blast resistant designs, earthquake bracing, and "crime prevention through environmental design" are other examples.

Regulatory Controls

Codes and standards are an example of regulatory controls. They serve to eliminate or reduce hazards to the built environment through land use permits (don't build in a flood plane or on unstable ground) and building codes that require minimum safety, construction, and engineering practices. Life safety codes define the required number of exits and specifications of evacuation stairwells. Automatic fire sprinklers are an example. Use caution when designing mitigation strategies based on these standards because they often represent minimum practices. Codes and standards are often revised after a disaster when engineers learn what went wrong with the previous version of the code.

Administrative Controls

Administrative controls include policy and procedure. A requirement that consistent platforms are used throughout the organization and a requirement that business continuity planning is included in all new project designs are examples. This is considered one of the least effective controls because it is dependent on human intervention. One company's policy failed when the roof of a building collapsed partially because the person assigned to measure and remove the snow accumulation was late because of the storm.

Service Agreements

Service agreements include contractual obligations for repair personnel to respond to an issue within a certain amount of time, the overnight replacement of damaged equipment or priority service in times of increased demand such as backup electrical generator fuel replenishment during a disaster or extended power outage. A common example is the subscription of a service to deliver a number of workstations preconfigured with your software image (that is, the standard programs you use) to your alternate work area within 24 hours of your need.

Redundancies/Divergence

In business continuity planning, the decentralization of facilities, systems, equipment, and processes goes a long way to help ensure the survival of the organization in times of a disaster. In today's business environment, many organizations believe it is in their financial interest to centralize as much of their processes as possible. Despite this, there are many opportunities to build redundant or diverse functionality into equipment and processes. Redundant array of inexpensive drives (RAID drives), dual power supplies and other critical system components, and uninterruptible power supplies (UPS) are examples of mitigation at an equipment level. A second, redundant data center and a data center configured on a load-balanced arrangement with diverse telecommunications routing are examples on a larger basis.

Separation of Hazards

Another form of mitigation involves keeping critical equipment, personnel, and processes away from hazards that can affect their functionality. Large water pipes running directly over the server farm or mainframe can be a hazard. Similarly, locating your critical processes next to the boiler room could become a problem if there is a fire or explosion. One company at risk for mail bombs located their chief executive's office just above the room where the mail was sorted. To mitigate these risks, one needs to be separated from the other.

Specific Mitigation

Some examples of specific mitigation include the following:

- Alternate power sources
- Alternate communications
- Policies and procedures
- Data backup
- Records management
- Facilities salvage and restoration

Alternate Power Sources.   Power surges, spikes, and drops account for the more common "disasters." These utility problems damage or destroy sensitive computer systems, research, and production equipment. Important data files can be corrupted. If power is lost, work in progress can be lost. Power losses over wide areas are expected to increase over the years as more demand is placed on the power grids.

The most common and least expensive mitigation to this risk is to install an individual UPS on each piece of critical equipment. A UPS is basically a device that delivers "conditioned" power (current protected from significant spikes or drops) to the equipment. It also contains a battery that, in case of a complete loss of power, will allow the unit to continue operation until a backup generator takes the load or the equipment can save its data and execute its shutdown routine. Some high-technology equipment can be damaged if it is simply shut off without going through this routine.

If the loss of electrical power will have a serious impact, consider bringing in redundant power from a different grid. This will prevent power losses caused by local conditions, such as lightning strikes or downed lines. Feed the power from a different direction and to a different part of the site.

Another common mitigation strategy is to install a backup generator capable of running critical (or emergency) systems as long as a fuel supply is available. If possible, have extra fuel on hand (if electrical power is out, the pumps at the local gas station will not work). Generators are powered by diesel, natural gas, or gasoline. Those supplied by natural gas might also be fed by separate sources or routes. Smaller generators mounted on trailers can be rented and brought on site. If this is the strategy selected, consider the installation of a "quick fit" device (transfer switch) outside the building. This device would be hardwired to the electrical distribution panel; the generator is simply plugged into the building, saving many hours of connection time.

Power supplies and generators should be tested regularly under load conditions.

Alternate Communications. Most organizations are highly dependent on communications for voice and data transmission. The loss of voice and data transmission can quickly have a severe impact on the organization. Equipment failure, software glitches, cable cuts, hackers, and fires in cable vaults or central stations can cause the loss of this function for days. Phone companies devote a tremendous amount of resources to ensure the reliability of their networks, but after a disaster, communications become the most quickly affected utility. Because of increased demand, the telephone network can become overloaded and cease to work, even if the equipment is undamaged. This can even happen internally, for instance, if a well-publicized event causes a sudden influx of calls they can overload the switchboard's ability to handle the traffic.

Strategies to mitigate damage to the communications system and to recover its function include the following:

- Service and replacement agreements
- Bypass circuits and fax lines
- Divergent routing
- Cellular backup
- Satellite systems
- Hot/cold sites
- Third-party call centers

When equipment fails or is damaged or destroyed, it will need replacement. However, many organizations cannot afford to be without communications for the time required to reorder, deliver, and install new systems. Most communications vendors offer 24-hour equipment replacement agreements for an up-front additional cost, with annual renewals of the agreement. If the impact of the loss of communications is severe or time dependent (such as in a catalogue sales operation), the organization may use a telecommunications hot or cold site. A cold site is a separate building or office area that does not have the equipment installed but has sufficient space to accommodate at least the minimum number of employees needed to conduct business. A call to the telephone company will transfer the company's lines to a hot site or third-party vendor where compatible equipment is installed and waiting to go so that the move will be transparent to the customers. Like hot or cold sites for computer systems, there are subscription, setup, declaration, and user fees involved. Modern call centers have operators on standby to answer questions or take messages until the firm is set up in the hot site. This initial switchover can take less than 15 minutes if the company is faced with a local (not regional) disaster.

Many systems are equipped with a number of "power failure" circuits that bypass the phone switch (your main on-site phone-switching equipment) and directly access an outside line. This capability, and the location of these circuits, should be confirmed, and used when necessary. If a fax server is not in use, phone lines for facsimile machines can be "borrowed" for voice or data communication. Check to see if your handsets are compatible (digital versus analog) with these circuits.

Voice and data are transmitted around the world through a variety of modes—overhead cable and fiber, underground cable and fiber, microwave, and satellite, to name a few. One of the more common causes of communications failure is a cable cut by a contractor digging a trench. Landslides and bridge collapses also disrupt communication cables. Diverse routing is one method used to protect against these dangers. With diverse routing, your main circuits may pass through southern states while your secondary or diverse circuits use cables located in northern states. Unfortunately, diverse routing can in reality mean only that your circuits use separate pipes—buried right next to each other. This can be true even if your primary and secondary cables are carried by different communications providers.

After the Loma Prieta earthquake in California, interest in cellular communications for emergency and recovery operations increased. Cellular systems minimize the use of ground-based cable, and the abundance of cell antennas adds redundancy to the system. Many believe they would be more survivable during and after a disaster. The switch to digital cellular makes data transmission and portable Internet access over the cellular system an acceptable strategy for a portion of the organization's recovery needs. The planner should use this strategy with caution, however; as in land-based systems, the increased number of users will congest the network even during nondisaster times. We see this occurring now in large metropolitan areas. After 9/11, newer forms of communication worked better than cellular, but these will suffer the same problems in the future. In fact, the U.S. government may approve the reallocation of cellular channels in an emergency to give governmental services more reliable use of the cellular network. In business continuity, in planning communications and telecommunications, diversity and redundancy are key to success. To add a degree of reliability, consider subscribing to two different providers, if this option is available in your area.

Microwave is a form of high-frequency radio transmission beamed from point to point. Microwave transmissions can be used to reestablish communications between buildings across a campus, city, or wider area. They can provide diversity in both voice and data communications, and they are not very susceptible to cable cuts. The use of microwave for diverse routing can be expensive, but it easily spans difficult terrain and provides large bandwidth capabilities.[4] Transmission towers, however, are susceptible to destruction or misalignment by high winds and earthquakes.

---

[4]The term *bandwidth* is used frequently when discussing continuity planning strategies for data and communication systems. It refers to the amount of information that can be effectively transmitted along a particular medium. It is measured by transmission speed. Think of bandwidth as a pipe: you can easily fit a ¼-inch dowel down a ½-inch pipe, but if you have a 1-inch dowel and only a ½-inch pipe, you need a bigger pipe, or you need to cut the dowel and send smaller pieces down the pipe. The same is true for data and telecommunications. You can only fit a certain-sized signal down a given size of wire at a certain speed.

Satellite transmission is used for primary or diverse routing, or as a backup communications channel. It is especially useful for continuity planning, in that mobile transmitters can be connected to the site and used to reestablish communications very quickly. Their failure rate is very low, and they are affected only by the difficulty with transportation after a disaster. Bandwidth and security are very good. Cellular phone companys, in some locations, are now using satellites as their cell antennas.

Third-party call centers and answering services, used for overflow customer support or order entry, can also be primary call centers after a disaster. Investigate the ability of their equipment and staff to handle the increased call volume. Consider sending some of your staff members who are familiar with the product and company to assist or train the center's staff.

Policies and Procedures.   Most "data" disasters are caused by people, not by equipment failure or natural catastrophes. Policies and procedures regarding data systems must be identified, implemented, and enforced to ensure a smooth and effective recovery. A data backup policy is an obvious starting point, but backup policies cannot be completely effective if the client does not store its data on the server. Most client/server architectures allow users to access two areas to store data—locally on the user's C-drive and also on the server. Although there are programs that back up the user's local drive (if the user's computer is running), this is a time-consuming operation and may back up unnecessary files. To avoid this, all users should store their data files on the server. Policies that restrict unauthorized installation of personal software or downloading material from the Internet will help to avoid disasters resulting from computer viruses.

Data Backup.   Most organizations can afford to lose a day or two of data or can tolerate the time required to reconstruct a small amount of lost work-in-progress. The primary strategy used in these organizations involves the nightly backup of each day's transactions onto magnetic tape. This is often done despite a lack of a policy establishing its need. Each organization must develop a data backup policy that requires the following:

- Nightly incremental backup of the server
- Weekly full backup
- Monthly archiving
- Yearly archiving

Backup tapes should ideally be taken each day off-site to a storage facility that specializes in the safe and secure storage of computer backup media. Offsite storage facilities should be audited annually to ensure the following:

- Good physical security
- That authorization lists are up-to-date and enforced
- That good fire prevention measures are in place
- That the building is structurally sound
- That it can find and deliver tapes and manuals in a reasonable time

If this is not possible, try to store tapes in a fire-resistant cabinet located in a separate building. During an evacuation, take the tapes (or whatever media used) with you, if this can be done safely. At the very least, backup tapes should be sent off-site weekly, and the previous weeks tapes returned and recycled.

Records Management. Many businesses that have not recovered after disasters have considered the loss of their business records the primary cause. Most regulations affecting business continuity planning refer to record retention and recovery. It should be apparent that a major focus of business continuity planning is the preservation and recovery of vital records. The insurance industry refers to these records as "important papers." The loss of customer, accounts receivable, and asset lists can severely damage future sales, cash flow, and insurance reimbursements. Also, failure to protect corporate records could in some instances bring criminal sanctions against management.

Records are identified as vital if they are important to the continued and future operation of the company. Their loss will have a severe impact or make it difficult to remain in operation. Some firms distinguish records as either vital or important, depending on how much they would be needed after a disaster. Vital records are those absolutely required to recover and restore operations; important, or key, documents are those that will reduce recovery time. Examples include the following:

- Customer lists
- Securities and stock records
- Corporate minutes
- Deeds
- Articles of incorporation
- Bylaws
- Other corporate financial records
- Leases
- Patents and trademarks
- License agreements
- Accounts receivable
- Banking statements
- Tax documents
- Treasury records
- Payroll and benefits information
- Research and development
- Information and specifications
- Insurance policies
- As-built drawings
- Clinical trial results
- Food and Drug Administration security files
- Food and Drug Administration filings that support regulatory requirements
- Business continuity plan
- Negotiations records
- Asset lists
- Laboratory notebooks

Once records are categorized, there are a number of options for safeguarding against their destruction, if they cannot be quickly re-created from their original source. The type of media they are stored on and the time dependency of the information may affect your recovery strategy. Most records are paper documents stored on-site. At the very least, they should be stored in locked, fire-resistant cabinets. The most important documents should be photocopied and stored off-site at a facility that specializes in document storage. As with data storage, always audit the security, fire safety, rapid retrieval, and environmental controls used by the storage company. Copies can also be stored at off-site locations within the organization. Document tracking and rapid retrieval can, however, become a problem when storage is internal.

Many records are copied onto microfiche, a technology that is becoming less popular with the increased use and lower costs of document scanning. Document scanning decreases retrieval time and reduces storage space. The documents are stored electronically. They can be sent directly to electronic vault facilities for storage and returned to recovery locations through the telephone system.

Once vital records are identified, it is important to implement policies and procedures that ensure their routine backup. Not all documents need backup; many can be re-created from the originals at vendors, customers, or regulatory agencies. Back up those documents that cannot be re-created without difficulty and those for which the delay required for their re-creation cannot be tolerated.

Facilities Salvage and Restoration.   A general recovery strategy that leads organizations to identify alternative processing and manufacturing facilities before a disaster, and to develop plans that allow for the rapid transfer of operations to these facilities, is typically very effective in getting back to some level of service. The goal of restoration is to return the organization to a predisaster state or to a defined strategic position. This cannot be done, however, in a temporary facility. The damaged facility must be rebuilt, relocated, or repaired. The rebuilding or permanent relocation of the organization is an issue to be addressed by the management team subsequent to a disaster. The planner will discuss these possibilities with management ahead of time and arrange for agreements and resources to expedite the search for new facilities—during conditions in which, inevitably, space would be scarce and prices inflated.

Most organizations return to their original facilities after a disaster, especially if the buildings are structurally sound. Before this can happen, the facility must be cleaned and made habitable. In the aftermath of a fire, there will be heat and soot damage to equipment. Smoke from a fire is acrid and corrodes sensitive electronic components, even after the fire is out. Water from sprinklers or firefighting hoses can damage documents and cause dangerous molds to grow. The drying and dehumidification of buildings requires specialized equipment and techniques. These services are provided by "restoration" companies, which either do the work themselves or subcontract to companies that specialize in the following areas:

- Salvage and debris removal
- Electronic component cleaning and repair
- Soot removal
- Dehumidification and drying
- Document drying and recovery

- Reconstruction (painting, plumbing, masonry, and drywall)
- Water extraction and moisture control

The safe recovery of vital records and books and the prevention of disease from water damage caused by flood, fire sprinklers, or roof collapse require special expertise and equipment. Restoration companies that specialize in a particular type of damage have expertise, experience, and equipment not generally available to internal facilities staff or to general contractors. Most general contractors can clean and rebuild a facility after a fire, but they may be unaware of the techniques used to eliminate the odor of smoke.

It can be very expensive to scrap damaged equipment, especially if it is unique or replacement lead times are extensive. Insurance policies may not pay the full replacement value of the damaged equipment. According to companies that specialize in the decontamination of equipment, restoration can save up to 75 percent over replacement costs, and restoration can be completed in a few weeks instead of the months potentially needed for replacement.

Most restoration companies will, for no charge, inventory assets and maintain construction plans off site. The use of these services can expedite the recovery process and allow management to concentrate on other recovery issues. In areas susceptible to earthquakes (that is, virtually every part of the United States), one of the first tasks necessary is the cosmetic repair of cracks and other damage. A large part of the psychological healing after a disaster is the return to normal surroundings. Employees staring at cracks in the wall are reminded of the disaster; restoration companies, if agreements are reached ahead of time, can erase the reminders that hinder a return to normalcy.

Cost-Effectiveness

Mitigation solutions must be cost-effective and technically feasible, and must not create additional hazards or problems. This should be obvious but must be considered. Mitigation is often a "big-ticket item." Be sure you have as much justification as possible before presenting your plan for approval and funding. You may need to prioritize your projects or spread them out over several budget cycles. Try to find solutions that can solve multiple hazard risks. Identify and cost-out alternative mitigation solutions to problems. This can help you achieve the most cost-effective measures or at least have others ready if management does not approve or fund your original plan.

One way to win approval for your mitigation plan is to demonstrate the amount of the loss your solutions will present (*losses avoided*). Provide a dollar value estimate of the structural, content, and displacement costs that would have occurred if the mitigation action were not taken. Use information from your business impact analysis to identify any potential loss of revenue. Include any maintenance and upkeep costs associated with the solution as well as any nonmonetary considerations, such as increased employee morale, customer satisfaction, or other subjective justifications. The losses avoided are most easily estimated for structural mitigation actions.

Displacement costs represent the dollar amount required to relocate a function, or a building's functions, to another location on a permanent or temporary basis. Consider moving costs, replacement costs if not included above, costs to prepare the new space, lost subleases, and other expenses as required. If moving to a temporary location, include the costs to move back to the original or to a permanent facility.

One possible shortcut to determine the expected future loss based on a past incident for which the previous impact is known is to multiply the impact (dollar loss) by the increase or decrease in exposure. Adjust the resulting figure for inflation.

A building or facility replacement cost is usually expressed in terms of cost per square foot and reflects the present-day replacement value. The cost may be offset by insurance, but be careful because that coverage may include only "present value" (depreciated value) and not total replacement. Include the cost to demolish a damaged facility, especially if there is some type of overriding expense, such as asbestos removal. Anticipating management's decision to replace a facility exactly as previously configured, or to use the opportunity for expansion, is probably best not attempted. Will repairs to the current building, if not replaced, require expensive building code upgrades? Projected repairs may be predicted by multiplying the replacement cost adjusted as necessary by the percentage of damage.

It may be advantageous or useful to understand the value of the functional loss of a building or facility. One method to make this determination is to add the budgets (or appropriate percentage of the budgets) or annual sales of the groups located within.

You may need to determine the content value of an entire facility if it is completely destroyed. Few businesses have complete inventories listed by location. The risk or insurance manager may be the best source of content data. Depending on your intended use (business impact analysis or mitigation analysis), you may need to use depreciated values or replacement values.

PREPAREDNESS

Before September 11, 2001, when we spoke of preparedness, it generally referred to the steps an individual or organization took to place it in a better position or to enable it to respond to and to survive the effects of a disaster. Unfortunately, we now need to include those steps necessary to prepare for a terrorist event.

Preparedness is having your plans and resources in place, keeping them updated, testing both the plan and those required to implement the plans. It gives you the capability to manage and respond to an incident.

In the United States, National Preparedness goals are established as a result of the Homeland Security Presidential Directives to develop capabilities to prevent, respond to, and recover from terrorist attacks, major disasters, and other emergencies. The directives also establish measurable targets, priorities, and methodologies for prevention and response to terrorist and natural threats. They are intended to guide federal, state, and local entities to determine how to devote limited resources to strengthen their preparedness efforts. The implementation of standardized response and management plans, collaboration and information sharing, and the strengthening of response capabilities are some of its goals.

Preparedness is an important step in the CEM cycle because materials needed to respond to an event must be in place and ready for utilization. When there's a fire, you don't call the purchasing department to order an extinguisher. Likewise, after a disaster, supplies and resources, as we have and will continue to mention, may be in short supply.

The more you have beforehand, the better your chances of survival. We most often think of preparedness as supplies and action, but it can also consist of knowledge. You need to know what to prepare for. Information and intelligence about the threat and its properties and consequences will enable a more rational and effective response. Preparedness can also include the following steps:

- Development of response procedures
- Design and installation of warning systems
- Travel advisories and employee tracking
- Establishing partnerships with local government (or with the business community if you are in government)
- Exercises
- Training
- Collecting personal information for kidnapping and ransom planning
- Stockpiling supplies and materials
- Entering into mutual aid or service-level agreements

Home and Personal Preparedness

Preparedness at home is important to the business or governmental operations because when workers know their families are safe, they are more apt to stay at work and deal with issues, or they are better able to respond back to work because they have less to deal with at home. Business and government should hold preparedness classes for their workers and provide them with a list of supplies and resources they will need. Basic topics can include the following:

- Emergency contacts and meeting locations
- Emergency supplies (food, water, medical, sanitary)
- Structural and nonstructural mitigation
- Insurance considerations and documentation
- Fire prevention and control
- First aid and cardiopulmonary resuscitation (CPR)
- Shelter-in-place instructions
- Emergency procedures
- Evacuation routes
- Warnings and media sources
- Light search and rescue
- What to do and not to do

Many local fire departments conduct Citizens Emergency Response Team (CERT or NERT) training to prepare residents for emergencies and disasters. Where available, consider sponsoring these programs for employees and their families.

Emergency Supplies

Many businesses also stockpile a 3-day or more cache of disaster supplies stored in containers located in the parking lot. The rationale is that the company may be responsible for those employees stranded at the workplace, or the supplies might be used for workers who need to remain to engage in recovery operations. They are also aware that after a disaster or terrorist event, these supplies may become scarce. Although this is a commendable practice, it can be expensive because the supplies need to be replenished every 3 to 5 years; also, locating all the supplies together exposes them to vulnerability to damage, theft, or sabotage, and cannot provide for the needs of individual employees. An alternate method, at least for smaller supplies, is to issue employees a carry bag of supplies (food, water, basic first aid supplies, light sticks, battery-powered radio, and so forth) when they are first hired. Employees then keep these in their desk or car with instructions to customize the contents to their individual needs (prescription medications, reading glasses, pictures of loved ones, clean underwear, and so on). The bag can contain the company or department logo. Specific guidelines and lists of recommended supplies are found on web sites of local government and other disaster or homeland security agencies.

Keeping spare parts or components such as access controller boards, RAID drives, or preconfigured laptop computers is another dimension of emergency supplies.

Public–Private Partnerships

Information sharing is important to business leaders in the midst of a crisis or emergency. By developing close relationships with governmental officials (emergency services, police, and federal agencies), you may gain better access to information and resources. Businesses may also have resources they can make available to government during times of need, not necessarily during a crisis or disaster. These partnerships can be difficult to foster because there can be an element of mistrust, regulation, and bureaucracy in the relationship between business and government. The advantages to both sides can be tremendous. Mutual planning and response, resource sharing, and information collaboration can make the difference in a successful response and recovery. These partnerships must be developed whenever possible; they allow you to make only a single telephone call to get what you need.

Service-Level Agreements

Vendor relationships are often cited as the most important element that gets organizations back in business quickly after a disaster. A simple phone call to the "right person" can get replacement delivery expedited. Companies are willing to pay up-front for agreements with vendors such as structural engineers, real estate brokers, contractors, and equipment rental firms to give them priority treatment. Some equipment manufacturers offer service agreements that provide for repair or replacement of preconfigured systems within 24 or 48 hours. It is also possible to subscribe to services that can deliver large numbers of servers and desktop systems preinstalled with your software and hardware configurations overnight.

Mutual aid agreements, or any arrangements you can make beforehand, will help to reduce costs and to ensure that things that are needed will have a better chance of presenting themselves for service.

Justification

Unlike mitigation, preparedness measures, although generally far less expensive, may be more difficult to justify or to illicit cooperation and participation. The threat of the disaster or attack must be seen as high in the short term, and the source predicting or advising of the threat must be seen as credible. Preparedness information is most effective if it is presented repeatedly through different media and in a form that is easy to use or to recall. Although many acknowledge the need, procrastination may delay implementation until it is too late. An interesting belief held by some is that once they have experienced a disaster and have survived, nothing worse can happen, or that they don't need further preparedness. Organizational preparedness may be more motivated by regulatory requirement or by the fear of litigation. It is harder to show a financial return for preparedness.

13 Response Planning

Eugene Tucker, CPP, CFE, CBCP
Contributing author

"Worrying about possible disasters can cause many sleepless nights for senior executives and board members, and prompt the creation of comprehensive contingency plans to help ease these worries. It is important, however, to maintain a delicate balance between over planning for events that may never happen and being adequately prepared to respond if disaster does strike."
—Jack E. Cox and Robert L. Barber, "Practical Contingency Planning," *Risk Management Magazine*, March 1996

Before September 11, 2001, most of our attention to emergency response was geared toward handling first aid issues or putting simple plans in place to meet regulatory requirements. Today, much of our attention is focused outside the organization to prepare for and respond to incidents that are terrorist acts. A good deal of concentration and resources are directed toward *first responders*—those people, usually in law enforcement, fire, and medical services, who would be the first to detect, report, and react to an attack by chemical, biological, or nuclear weapons.

Businesses are now incorporating measures to mitigate the effects of explosive devices, or an aerosol introduced into their ventilation system. Many emergency response plans now include instructions to *shelter-in-place*. Organizations that did not understand today's risks and that did not incorporate these new threats into their plan were some of the many who succumbed to the intent of terrorists to cause fear and overreaction by triggering, for example, a full-scale response by local officials, dressed in level A Hazmat suits, to clean up spilled artificial sweetener on café tables, fearing that someone had introduced anthrax into their facility.

The manner in which an organization reacts to an emergency is very visible and can be subject to much criticism, both internally and externally, especially if the emergency is a newsworthy event. A poor or nonexistent response can set the stage for a loss of credibility for the department, leading to legal action by stockholders and victims. Effective action taken to control an emergency will reduce injuries, protect assets and mitigate their loss, and position the organization for a smooth and rapid recovery—all good reasons why response is an important element of the comprehensive emergency management (CEM) model.

Organizations must be able to respond to situations they can reasonably anticipate during normal conditions and after disasters. Emergency services provided by local jurisdictions (police, fire, ambulance, and hospitals) may not always be available. When the demand for these resources must be prioritized after an area-wide event, businesses are virtually always at the bottom of the list; government believes that the business community has the resources to be self-sufficient and that other organizations, such as public schools, are more in need of its services.

In the United States, an emergency response capability is required by federal and many state regulations for all organizations. All employers must maintain emergency action and fire prevention plans. The plan must be in writing if there are more than 10 employees. The size and type of the plan and of the response capability is dependent on the type of business (for example, hazardous materials producers have much higher requirements than do general businesses). As with most plans, they must be simple to follow and the necessary guidelines easy to find. Unnecessary verbiage should be excluded from instructions used in the field.

State and local jurisdictions may impose additional requirements on the business community to develop a response capability. These requirements supersede federal standards if the local regulations are more stringent. The National Fire Protection Association (NFPA) Section 1600 (2004), entitled "Standard On Disaster/Emergency Management and Business Continuity Programs," outlines a minimum standard for the design and implementation of both response and recovery programs. If adopted by the local authorities, you may be responsible for its contents. The National Commission on Terrorist Attacks Upon the United States, better known as the 9/11 Commission, recommended that NFPA 1600 be recognized as the National Preparedness Standard. The Private Sector Preparedness Act of 2004 (HR 4830) adopted most of the elements contained in NFPA 1600.

Many of these regulations are matters of law and industry standards; accordingly, managers can be held liable in civil and criminal court if the programs they require are not implemented. Additionally, violation of these laws is excluded from liability insurance coverage for "errors and omissions" and "directors and officers." Civil courts are increasingly holding companies liable for not implementing adequate emergency response plans that protect the well-being of their employees and guests.

Emergency response is actions taken to manage, control, or mitigate the immediate effects of an incident. An emergency response plan is not a disaster recovery or business continuity plan, a mistake often made by business managers.

The United States has established the National Response Plan (NRP), which specifies how the resources of the Federal Government will work in concert with state, local, and tribal governments, and the private sector to respond to "incidents of national significance." Its purpose is to establish a comprehensive, national, all-hazards approach to domestic incident management across a spectrum of activities, including prevention, preparedness, response, and recovery. The NRP is predicated on the National Incident Management System (NIMS). NIMS, under development at the time of this writing, consists mainly of a management system (the Incident Command System), preparedness (planning, training, exercising, certification, equipment acquisition, and publications management), communications and information management, and supporting technologies. Together, the NRP and the NIMS provide a nationwide template for working together to prevent or respond to threats and incidents regardless of cause, size, or complexity.

Significant elements of these standards and regulations include the use of the Incident Command System (ICS), hazard-based planning (multihazard functional planning; see Chapter 16), and coordinated response (unified command).

THE INCIDENT COMMAND SYSTEM

The ICS is a hierarchical management structure used by governmental agencies, fire, and police to respond to an emergency. It is primarily a field response management system but is adapted for use in the Emergency Operations Center (EOC). Devised by the fire service in 1971, it provides guidelines for common multiagency operating procedures, terminology, communications, and management. Its modular structure allows for a consistent and coordinated response to incidents of all types and complexity. Because of the growing interdependency among the response organizations of industry, business, and governmental agencies, the use of the ICS by business and industry is necessary. Emergency response teams (ERTs) may already be required by law or industry standard to utilize this system. If the organization's response requirements don't warrant the use of an ERT, business owners and responsible managers (including recovery planners) must still be aware of the methods and protocols, such as ICS, that are used by the local jurisdictions to manage emergencies.

ICS is a management tool that relies heavily on the concept of management by objectives (MBO). Response objectives are set by the senior responder and delegated to the subordinate positions after agreement that the objectives can be met. The senior responder is referred to as the *incident commander*. By using this approach, the incident commander can coordinate the response to complex and technical incidents without unreasonable expectations. ICS is also sensitive to the basic management principal of span of control that limits the ratio of subordinates. If the incident is small and the response is relatively simple, the ratio is eight subordinates to one. If the crisis expands and becomes more complex, the span of control is reduced to provide the most effective leadership. Some believe the span of control in an emergency situation should be five subordinates or less. ICS is divided into five major functional units. The fire service version is expandable to 36 positions, but most are not relevant to business response. The five units or sections are as follows (Figure 13.1):

- Incident command
- Operations
- Planning and intelligence
- Logistics
- Finance and administration

With small incidents, it is not necessary to establish all units (sections) of the ICS. In this case, the incident commander (see subsequent section) will directly manage or assume the duties of each of the units or activate the units as additional personnel arrive. Operational need is the primary factor in determining what is activated.[1] Each unit is

---

[1]ICS for executives: *Standard Emergency Management System Executive Course, Student Reference Manual,* State of California, August 1995.

FIGURE 13.1 Basic Incident Command System Structure.

headed by a section chief and may be further divided into subsections as required by the complexity of the incident or need to maintain the proper span of control.[2]

Incident Commander

The incident commander has overall responsibility at the incident or event. A distinctive vest that contains the words "Incident Commander" is worn as identification. The incident commander determines objectives and establishes priorities based on the nature of the incident, available resources, and agency (company) policy.[3] The role of the incident commander is usually filled by the first responder to arrive at the scene, who is relieved of this duty when a more senior responder or a designated incident commander arrives. A *command post* is set up at a safe distance near the location of the emergency where the incident commander will manage the response. Once established, the command post should not be moved. It can be located in the field, at a vehicle, inside an office, or where reliable communications (electronic and verbal) and security (access control) can be maintained. When appropriate, it should be within view of the incident, but away from noise associated with the incident.

Management must delegate (ahead of time) to the incident commander the authority to make the tactical decisions necessary to stabilize or end the emergency without interference. Management's role is in the EOC to make strategic decisions based on the events or to allocate resources among multiple incidents. Reliable communications between the EOC and incident commander are essential. The incident commander follows preexisting policy set by management and will use standard forms and checklists to ensure that all tasks are completed. Software programs are available to aid in the management of the emergency.

Some of the specific duties of the incident commander are as follows:

- Overall field management of the emergency
- Coordination with the EOC or other incident commanders. The incident commander of the firm's ERT should colocate with the fire or police department incident commander (an element of Unified Command).

---

[2]The Law Enforcement version of ICS uses the term "Officer-in-Charge" or "OIC."
[3]Ibid.

- Ultimate responsibility for the safety of responders
- Approval of all plans and resources
- Situational analysis
- Setting objectives and priorities
- Delegating authority as necessary
- Primary responder until others arrive

If the size of the emergency warrants the establishment of the following positions, assistants to the incident commander include an information officer, safety officer, and liaison officer.

Information Officer

The information officer, or public information officer (PIO), is the news media contact for the event. In a business environment, the public relations representative will fill this role and should be less (or not at all) subordinate to the incident commander, as he or she would be under the government's version.

Safety Officer

The safety officer monitors safety conditions, ensures that compliance regulations are met, and develops measures to ensure the safety of all assigned personnel.[4] The safety officer is often responsible for evaluating changing conditions and should have the authority to withdraw responders or to suspend an operation without clearance from the incident commander.

Liaison Officer

The liaison officer assists the incident commander on larger incidents to which representatives from other agencies may respond by coordinating their involvement and providing them with information on conditions, objectives, and resources.

Operations

The operations section implements the action plans and objectives issued by the incident commander. These are the "doers" of the response. They participate in the selection and reality-checking of goals and direct all resources necessary to carry out the response. A constant flow of situational information and milestone achievement is communicated back to the incident commander. Operations can be subdivided into functional or geographical divisions as needed. Examples include first aid, search and rescue, and Hazmat cleanup.

Planning and Intelligence

The planning and intelligence section develops the incident action plans to implement the goals and objectives of the incident commander. As part of their plans, this section

---

[4]Ibid.

also determines what resources are needed to accomplish each task. Members of this section must gather information about the incident before they can devise a meaningful plan. In a large-scale incident, this section will accomplish the following:

- Collect intelligence (analyze conditions and the scope of the incident)
- Project or predict changing conditions
- Prepare action plans
- Prepare contingency plans if conditions, events, or resources change
- Track resources available, in service, and used

Technical advisors are included in the planning section to provide expert advice when needed. Chemists, safety engineers, toxicologists, industrial hygienists, meteorologists, and structural engineers are examples of the types of experts that might be included in the response.

Logistics

The logistics section obtains all resources and services needed to manage the incident. This section delivers personnel, equipment, food and supplies, restroom and shower facilities, and so forth. The logistics section simply supplies resources. The planning section is responsible for resource management and use.

Finance and Administration

The finance and administration section maintains records and documents the history of the response. It projects, tracks, and approves expenditures by the logistics section, and completes a final cost analysis of the response. Documentation of times, events, and actions is important to the postincident analysis, insurance reimbursement, criminal prosecution, and defense of a civil action.

Example

This is how it might work—the incident commander might issue the command (goal) to extinguish the fire. The planning section determines that the fire is small in origin, involves general combustibles, and will require dousing with water by one hose team. If no hose is available, logistics will find one. The operations section will then grab the hose and put out the fire. This is noted by the incident commander and the planning section personnel, who are also responsible for tracking any resources used. The cost of the hose, the damage caused by the fire, and the time and other expenses used to put out the fire are tracked and reported by finance and administration.

UNIFIED COMMAND

Unified Command is used when multiple agencies or jurisdictions are involved in the response to an incident. Its structure is that of the ICS, but unified command planning combines the objectives of all incident commanders so that the overall response can operate as if it were a single-agency incident, avoiding working at cross-purposes to other agencies. Resources are shared, but participating agencies do not have the authority to approve or disapprove the objectives of others. In the response to certain types of events, such as a terrorist bombing, certain agencies will take a lead role and therefore act as incident commanders. They may form a Joint Operations Center (JOC).

Despite of the necessity of Unified Command to coordinate the response to a large incident, it is often difficult to implement because of the territorial or jurisdictional battles that still exist in today's response community. Frequently, exercises that incorporate the use of Unified Command, especially those that involve public–private interface, are not completed as often as necessary. Public agencies and the business community must recognize the need to work together.

EMERGENCY OPERATIONS CENTER

The EOC is a location where the management team members or emergency managers meet to direct or coordinate the response to a large-scale incident or to begin the direction of a business's recovery. To the business continuity planner, the EOC or command center is where management coordinates the actions of the individual recovery teams, monitors the progress of the recovery, and passes requests and information up, down, and across the structure of the recovery organization. Although its primary function is strategic, the EOC can make such tactical decisions as the allocation of resources between competing teams, incidents, or plans.

The organization must anticipate the complexity of its response and recovery and design an EOC that accommodates its operational requirements. From the business perspective, the physical and operational layout can be as simple as a conference or hotel room. Companies with numerous sites in different geographical areas need (as do governmental agencies) to design theater-style complexes that have adjoining rooms and integrated audiovisual, computer network, and communications systems and support personnel. The EOC should be located in a secure, structurally safe building that is centrally located and easily accessible to the responders even when transportation systems are disrupted. An alternate EOC, such as a mobile trailer, located a distance from the main site, should always be available.

The EOC centralizes control of disaster response among individual recovery teams or among multiple governmental agencies. Large EOCs can utilize an ICS structure internally and actually configure the seating or work space along functional lines.

The primary functions of a governmental EOC are as follows:

- Coordinate the response to large or multiple events
- Create or refine policy
- Allocate resources
- Collect and manage information about the incident, responses, and decisions
- Release information to the public
- Maintain appropriate records

For most businesses, the EOC is a large conference room that has a sufficient number of phone jacks, room for status boards, and separate work space for the management team leader and recovery coordinator. The phone jacks should be wired to allow the firm's phone switch to be bypassed in case of power failure or the destruction of communications equipment. Extra phone sets and fax machines, as well as all supplies needed to operate the EOC (radios, extra batteries, overhead projectors, whiteboards, forms, and office supplies), should be stored in the room or close by. Duplicate supplies and equipment should be stored in the alternate EOC if possible.

The flow of information, both from outside and within the EOC, is critical to the decisions made during a crisis, especially in a large-scale operation. Technology, such as computer networks and multimedia displays with colorful graphics, will speed the delivery of accurate information to decision makers and will greatly reduce fatigue. There are several software programs on the market that are designed for use in an EOC and can help to satisfy these requirements. External information must flow between the incident commander (or recovery team leaders) and the EOC. Situational or conditional reports must be available. The EOC must have the ability to communicate with the outside world even when power and normal modes of communication are disrupted. Backup (redundant) communications, such as cellular or satellite telephones and amateur radio, should be utilized.

Other considerations for the EOC include the following:

- Don't overwork the EOC staff. Decisions made under stressful conditions are often less effective. Add long hours to the stress, and the quality of the decisions can deteriorate even more. Shifts should not exceed 12 hours. Arrange for rest periods, breaks, professional massages, and plenty of food and water.
- Access control (security) is important, especially in a large EOC. Unauthorized visitors, media personnel, and managers not directly involved in the EOC operations must be denied access to the center. Consider a badging system to help control access to the EOC.
- A video conferencing capability is useful, as is an ability to monitor ATV (amateur television) broadcasts. In a recent California flood, the local ATV club rented a helicopter and sent live video back to an EOC.
- All EOC operations, rooms, and equipment should be connected to backup power generators.
- The EOC must be "user friendly" with respect to both comfort and functionality. Poor lighting, high noise levels, difficult-to-read visuals, poor ergonomics, and other negative "human factors" will tend to fatigue the staff sooner and adversely affect their ability to make intelligent decisions.
- The EOC must be designed to support multishift staffing and to operate continuously for extended periods.
- Keep operations as quiet as possible. Establish separate meeting rooms and a soundproof radio room; deliver television audio through headphones, and use telephones that light instead of ring.

EMERGENCY RESPONSE TEAM

An ERT is an internal organization of employees designed to respond to emergencies before the arrival of public agencies. Although membership in the team is usually voluntary, employees from certain departments are generally necessary to ensure an effective response. This includes security (access control, communication), facilities (resources, equipment shutdown), and environmental health and safety (technical support).

An emergency response team can help the company to accomplish the following:

- Intervene and stabilize emergencies with less delay
- Reduce injuries and loss
- Prevent adverse publicity
- Demonstrate management concern and support for the safety of employees
- Minimize the impact on the environment
- Help comply with regulatory requirements to mitigate hazardous materials incidents (Occupational Safety and Health Administration, Uniform Fire Code, Environmental Protection Agency)
- Become the sole response in a disaster situation when public agencies do not have the ability to respond

Some basic steps required to form an ERT include the following:

Management Acceptance and Support

The effective and safe operation of an ERT involves a large commitment of time, money, and resources for training and equipment. The time required for training even a small ERT in a low-hazard environment should be no less than 8 hours each quarter—much more in a larger, risk-intensive environment. The type of hazards the team may respond to will also affect the minimum training and equipment requirements. Teams involved in higher levels of "hazwoper" (hazardous waste operations) require a minimum of 40 hours of training. This specialized training and equipment, and the "lost" time of members away from their normal duties, can add up to a significant expense for most organizations.

Duties and Responsibilities

What is the scope of the team's duties and responsibilities—that is, to what type of incidents are they expected to respond? This will often define recruitment standards, equipment requirements, and training levels. Medical examinations and inoculations may be required.

Planning

Develop response plans, team structure (conforming to the ICS), policies, and procedures.

Determine Equipment and Resource Needs

Budget for short-term and long-term requirements. Equipment may include colored vests, hard hats, spill-kits, and a list of other items. Don't forget to include communications equipment. Two-way radios, pagers, and police and fire scanners form a basic response-communications system.

Recruit Team Members

Include sufficient numbers to field full teams for all shifts, to meet the staffing demands of each type of incident, and to allow for a reserve force to substitute for sick or injured team members. Provide incentives to create interest.

Develop Training Programs

The training must fit the scope of the team's duties and regulatory requirements. It may be necessary to bring in outside consultants or to have the local fire department assist with training.

Conduct Regular Drills

Practice, practice, practice. Arrange joint drills with public agencies.

Advertise

Let employees and the community know about the capabilities of the team. This will increase confidence in the organization's ability to respond to emergencies, and it should elicit better cooperation from employees.

EMERGENCY PROCEDURES

It is necessary to identify and understand the characteristics of hazards so that the planner is in the best position to prepare for, mitigate, and control their consequences. Without this understanding, we may face unwelcome surprises during the response phase, and needlessly spend valuable time resolving unanticipated issues during the recovery and continuity phase. Hazards can produce different effects depending on their magnitude, their duration, their time of year, building construction, and the level of preparedness and mitigation. On a regional basis, hazard impacts can vary based on demographics, location, building habits, and conditions that may increase or reduce the effects of the hazards (soil conditions that may cause liquefaction, creek or river debris, diseased forests, and so on). Be aware that one type of hazard can produce multiple hazards, so-called secondary hazards. Hurricanes can generate tornados and flooding, flooding can cause fires and bridge washouts, and earthquakes can cause Hazmat incidents and tsunamis.

The following sections provide guidelines the reader may use to plan a response to emergencies. The guidelines are generic and are not intended to list all foreseeable emergencies, or to list all measures the organization should take to prepare for, respond to, and recover from an incident, natural hazard, or disaster. They are primarily directed to industrial sites and to business in general and do not consider situations in which the activation and notification of other agencies or entities may be required. Many of the guidelines and conditions apply to multiple types of disasters, although they are listed here only with the section with which they are most commonly associated. In all cases, always assess the situation before formulating a response.

Many emergency response instructions begin with the admonition "do not panic" or "prevent panic." People rarely panic before, during, or after an emergency, except under very specific circumstances. Assigning responsibility to someone to prevent panic is a task that few, in any, can accomplish if panic were to develop. This instruction should not be included in emergency response plans. Also not included in the guidelines listed in this chapter are methods to activate emergency responders and, except in some cases, the need to notify others within the organization. Planners should identify these needs and document them appropriately.

BOMB INCIDENT MANAGEMENT

The days since 9/11 have shown that terrorists are capable of using nonconventional means to carry out their objectives. In our rush to protect against these new forms of attack, we must not lose sight of the more conventional attack—the simple, tried, and true tool of both domestic and international terrorism: bombings.

Partially to distinguish between devices manufactured for the military, terrorist bombs are referred to as *improvised explosive devices* (IEDs). Although almost all the incidents within the United States and Canada are rather small scale (with a few, very notable exceptions), transportation systems and other critical infrastructure, oil refineries, businesses with hazardous materials, public assembly, and other operations that could cause collateral or synergistic damage are prime targets. Many believe the types of devices used, most commonly pipe bombs and incendiary devices, will expand into a greater use of package bombs, greater-yield automobile and truck bombs (even ambulances have been used to hide explosives), and "dirty bombs" carried by suicide or homicide bombers. Large trucks have been used by terrorists as vehicle-borne improvised explosive devices (VBIEDs) with great success.

Throughout history and throughout the world, bombings are a prime tool used by terrorists and are expected to remain the dominant threat. Materials are easy to obtain, terrorist groups have a long history of expertise in their use, and the terrorist can be a continent away when the bomb is detonated. In 1996, there were 1,685 bombings and attempted bombings in the United States that caused $2,260,362 in damage and 24 fatalities. In the same year, incendiary incidents totaled 533 and $7,058,368 in damage. Seven people died from these firebombs. Between 1993 and 1997, bombing and incendiary incidents in the United States totaled 13,560, killed 478, and caused $645,947,792 in damage. Letter bombs (package bombs), depending on the year examined, typically represent less than 1 percent of the above totals. According to the Bureau of Alcohol, Tobacco, Firearms and Explosives (ATF), total incendiary and explosive incidents decreased in 2003 to a total of 386, with 7 fatalities and almost $5.5 million in damage.

The ATF also reported that in 1996, 2,983 pounds of high explosives were stolen; between 1992 and 1996, this number was 27,562. Interestingly, the ATF reported that 27,562 pounds (1996) and 93,278 pounds (1992–1996) were recovered.

Vandalism consistently leads as the motivation of bombers (mailboxes are a common target), followed by revenge (sour love affairs, ex-employees, and "messages" to our beloved Internal Revenue Service, for example). Homicide, suicide, protest, extortion, insurance fraud, and labor disputes are near the bottom of the list.

The Oklahoma City bombing consisted of a 4,800-pound ammonium nitrate and fuel oil mixture that generated a blast pressure of almost 6,000 pounds per square inch and created a 30-foot wide, 8-foot deep hole. The winds spawned to replace the vacuum caused by the explosion exceeded 1,000 times hurricane force. Windows were blown out 1,000 feet away. Even in a small-scale explosion, structural damage is possible. Nonstructural damage can include fire and smoke damage, water damage, soot and dust, glass, and debris. Electronic equipment and circuit boards can be damaged from the blast, and the psychological effects can be significant. A bombing or the discovery of a device can put the firm on legal notice that such incidents are foreseeable and thus increase future liability.

Threat Evaluation

In the United States, the chance of finding a bomb after a threat is very low. The level of physical security and the degree in which it is enforced are primary considerations when it comes time to evaluate the credibility of and the response to a bomb threat. Although most security professionals no longer recommend the evacuation of a facility after most threats, a search should be conducted in all cases.

Threats can be emailed, snail-mailed, or most commonly, telephoned into the business. It is important to remember that most telephoned threats are hoaxes. Only 2 to 6 percent or less of bombings are preceded by a threat or warning. There is some variation in these numbers, but some experts put the probability at 0.5% (1 in 200 chance), whereas others, such as the San Jose, California Police bomb squad, put the odds at "infinitesimal."

Generally, the purpose of a valid threat is to prevent injury or to prove intent in an extortion case. In these instances, familiarity of the site and the increased detail provided by the caller adds credibility. The caller may provide the exact location, time of detonation, and a detailed description of the explosive. Although "bomb threat checklists" are found at many reception stations, it is important for someone in your organization to develop the capability to evaluate the credibility of any threats received. This capability can reside with the security director, risk manager, crisis management team, or other responsible person who can be reached quickly and has the ability to make rapid decisions. Most police officers and police dispatchers are not trained to evaluate threats. Many are restricted by policy to not offer advice to the business owner. The person or people who evaluate the threat must have some training in threat evaluation that includes an understanding of the firm's threat risk profile. Basic components of the profile are the following:

- Level of security
- Evaluation of the visibility of the company or controversial business activity
- Recent events (product recall, pending labor actions, reduction in force)
- History of bombings and threats
- Activity of individuals and groups (intelligence)

The wording of the threat is examined for key words ("device" as opposed to "bomb," terms that describe the firing train, and so forth), conditions and circumstances (background noise of a party may indicate a low credibility, or a day in history that is significant to a suspect group might indicate a higher credibility), detail of the threat, ability to carry out the threat (security level and profile), and motivation. The results of the evaluation will guide the decision to evacuate, shelter-in-place, close the business, or to take other action as appropriate.

Evacuation

The decision to evacuate the building or site after a bomb threat is still a controversial and difficult action, although many security managers now follow the philosophy of avoiding evacuations unless a device is actually found. There is good reason for this approach: because most threats are false, evacuation will cause needless loss of productivity, decrease employee morale, increase the possibility of injury from the evacuation, and most often, satisfy the intent of the caller. Once a threat has been communicated to employees through an evacuation, expect additional threats generated internally. Most bombs in business settings are placed in the parking lot, building perimeter, or public areas, such as the lobby, hallway, and lockers. Evacuating employees will send them through and to these areas. If the threat is not credible, begin a search. If nothing is found, return to normal operations. If a device is found, evacuate a safe distance away from blast effects, or send employees home. If the threat is credible and a detonation time of less than 30 minutes is given, some may want to delay the search and evacuate if the employees can be cleared from the building and placed in a safe location before detonation. If the threat is credible and a detonation time is greater than 30 minutes, initiate a search if it can be completed in time to safely evacuate afterward. These times are somewhat arbitrary and should be adjusted according to your circumstances. This criterion applies only to threats received in the United States. The characteristics of threats and groups overseas may require a different approach.

Searches

Although you should rarely evacuate after a threat is received, you should almost always begin a search for a bomb, even if the threat is obviously a hoax. Searches should be conducted internally because most police officers are not trained in search methods or are not allowed to conduct a search of your premises. Calling the police to report a threat or to conduct a search will often result in the incident becoming part of the public record, open to publication by the media.

Searches can be conducted by the security force, ERT, or facilities department in conjunction with department managers, or by a combination of these methods. Bomb sniffing dogs are very effective, but their extended response time often makes their utilization impractical. Training in search methods and bomb recognition is desired but not overly important because a bomb can look like anything; any object that is out of place, such as an unidentified, unclaimed briefcase, a package next to the gas main, or a small pipe sitting outside the data center, is suspect. It is most important to identify search methods and searchers ahead of time.

Any search must be systematic, rapid, and thorough. Sectionalize portions of the site and building and prioritize the order in which these sections are searched. Assign specific search areas if they can be searched concurrently, or assign the sections as personnel become available. A good starting place is any area mentioned in the threat (I always check the bottom of my chair first). Search next the most common, accessible areas bombs are found (building perimeter, public areas), then areas that could provide a synergistic effect (hydrogen tank, gas main), and then areas critical to the business operation (backup electrical generator or data center). Search last the noncritical areas or areas difficult to reach. Assign these sections to teams most familiar with the areas, such as department managers, who have the advantage of increased search speed, a greater likelihood to recognize objects out of place, and a better knowledge of critical equipment and hiding places.

Rapid communication when conducting a coordinated search is vital. Most bomb incident procedures carry the admonition to turn off all radio transmission devices, including cellular phones, based on the fear that using your radio or phone will cause the premature detonation of a device. Although you may not want to transmit adjacent to a suspicious object, the chance of a predetonation from your radio is remote. Nevertheless, you should minimize the use of portable communication devices.

Suspicious Object

If a suspicious object is located, try to find its owner. Isolate (secure) the area from entry. Evacuate people away from the area or send them home. Remember that the area is now a crime scene and that the facility may be closed for 24 to 36 hours or more. Inform the appropriate emergency personnel (police, fire). If safe, open doors and windows around the blast area and shut off hazardous processes in the area. Consider shutting off utilities. If safe, continue to search for other devices. Do not touch or move the object (even if it is a dud). Do not cover the object or cut any wires. Do not put the object in water or pour water on it. Activate your crisis management plan.

Package Bombs

Explosives received through the mail (mail bombs) can be envelopes, boxes, or packages and may have one or more of the following characteristics:

- Unusual postmarks or places of origin
- Excessive postage
- Incorrect addresses or titles of recipients
- Excessive handling, wrapping, taping, or inappropriate bulkiness
- Excess weight, stiffness, bulges, or uneven balance and feel
- Smudges and greasy-looking spots or areas
- An odor of almonds or a chemical odor
- Protruding wire or string
- Pinholes from which a safety-pin arming device may have been withdrawn

The use of nonmetallic letter openers can help to prevent a detonation, but if there is any doubt, isolate the letter and have it examined. Inexpensive sprays exist to quickly identify any explosive residue on the package. If you are a high-risk enterprise, consider small x-ray devices in your mail room.

Suicide Bombs

Robert Mueller, the current Director of the Federal Bureau of Investigation, believes that suicide bombers are inevitable in the United States. Powerful enough to kill almost everyone in a 50-foot radius, they can destroy busses, restaurants, and trains. Suicide bombings are a favorite and effective method of terrorists to penetrate a target. According to former Central Intelligence Agency (CIA) Counterterrorism Chief Vince Cannistraro, "There is no 100% defense against suicide bombers."

These devices usually consist of 10 to 30 pounds of explosives and are strapped to the body, hidden in clothing or other objects like backpacks. Suicide bombers may also be the drivers of cars or trucks laden with explosives who detonate their cargo over a bridge or near a crowded intersection. Suicide bombers can be difficult to identify because they no longer fit the profile of the past (perpetrators were almost exclusively young men, but women and children are now found to have joined the ranks). Potential indicators include a person who is alone or with one other person and may appear nervous, apprehensive, or agitated, wearing loose or bulky clothing, possibly inappropriate for the weather (overdressed for summer heat). Exposed wires may be visible, and their midsection may seem rigid. Their hands may be clenched around a trigger (a second person may trigger the device with a call to a cellular phone).

In the case of a detonation, first responders should not approach the suspect or the suspect's remains. There may be undetonated or partially detonated explosives or secondary devices present. A bomb squad should determine when it is safe to approach. If the incident occurred within close proximity to your location, consider taking initial steps to protect against dispersion of radiation, chemical, or biological agents.

Dirty Bombs

IEDs specifically designed to cause injury, such as a pipe bomb, can be filled with nails, pesticides, or other poisons to cause the maximum amount of injury. The combination of radioactive materials with such an explosive device produces a "dirty bomb," more properly called a radiological dispersal device (RDD). The intent is to use the force of the explosion to distribute radioactive material throughout the area. Waste byproducts from nuclear reactors, medical devices, or other radioactive sources are the likely materials found in these devices. This combination of explosives and the radioactive material available to put into the device cannot produce a nuclear explosion. If dispersed through the air, the resulting contamination could affect several city blocks, but based on the probable type of material available to a terrorist, would not significantly cause severe illness, a marked increase in cancer cases, or deaths from exposure to radiation unless the device is very large and uses a credible radiation source, and the victims are very close to it. This is assuming that multiple devices are not detonated in unison. Inhaling radioactive dust from an exploding RDD could represent a health risk. The extent of contamination depends on the size of the explosion, the type and amount of radioactive material, and the local weather conditions at the time of the explosion.

Effective RDDs are difficult to acquire and to make. The Iraqi government and other military forces have tried and failed in the past to produce a dirty bomb capable of lethal results apart from the initial explosion; although in 1996, Islamic rebels from the break-away province of Chechnya planted, but did not detonate, such a device in Moscow's Izmailovo Park to demonstrate Russia's vulnerability. Today, only the United States and the Soviet Union have manufactured portable nuclear devices small enough to fit into a large backpack. The existence of "suitcase bombs" that contain nuclear devices manufactured in the Soviet Union is believed by many to be a myth. Many large cities and port facilities, however, have installed radiation detectors to identify such devices.

Apart from deaths caused by the explosion, the greatest impact from an RDD is the costly cleanup and the lingering fear of reentering the exposed area or building. Dirty bombs are primarily designed for fear and intimidation. An example of their effectiveness is the rush of people who purchased potassium iodide (KI) after the early reports of the potential use of dirty bombs in the United States. Unfortunately KI will not protect from the effects of an RDD unless the device contains large quantities of iodine isotopes. KI protects the thyroid gland from only radioactive iodine; it offers no protection to other parts of the body or against other types of radioactive isotopes. There is no guarantee that these isotopes would be used in the device.

Mitigation

The type of mitigation used should be appropriate to your risk. Consider the following:

1. Develop an intelligence program (especially if your risk is from domestic terrorist groups) and maintain continuous contact with law enforcement.
2. Complete an engineering evaluation of the blast resistance of your facilities and implement measures to mitigate the effects of a detonation by creating a blast-resistant exterior and structural elements that can resist blast loads. Explore the use of computer-based modeling and decision support systems to assess the extent of blast damage to a building's structural frame. When designing new construction, incorporate blast-resistant architecture.
3. Locate ventilation systems away from ground-level entrances, parking areas, and streets. If possible, locate them on the roof. Install automatic dampers to close air intakes when not in use.
4. Ensure that fire prevention and protection systems are up to the latest codes and best practices (an expensive proposition). Use blast-resistant separation for fire system equipment, pipes, and controls.
5. Locate critical facilities, offices, and processes away from the perimeter of a campus. Within a building, locate them away from parking lots and loading docks, and away from the mailroom. If justified by the risk, consider locating the mailroom in a separate building. The president's office of a company that was identified as a potential target of the "Unabomber," who specialized in package bombs, was located one floor directly above the mailroom.
6. Remove hiding places for IEDs from around the building perimeter and especially near lobbies, emergency exits, or paths evacuees may take. This includes trash receptacles, mailboxes, large planter boxes and hedges, newspaper racks, and so forth.
7. Design streets, campuses, and approaches to buildings to prevent vehicular traffic from having a straight approach to a security checkpoint or to building lobbies. This will preclude vehicles from reaching high rates of speed and crashing through the checkpoint or

into building lobbies. Place fountains, speed bumps, bollards, sculptures, or other objects in front of lobbies.

8. Control access to parking areas, especially if underground. Physically inspect vans and other large vehicles; check the undercarriage of vehicles, under the hood, and in the trunk. Provide vehicle inspection training to security personnel.

9. Establish no-parking zones around facilities and enforce with tow-away procedures. Approach all illegally parked vehicles in and around facilities, question drivers, and direct them to move immediately; if the owner cannot be identified, have the vehicle towed by law enforcement.

10. Isolate employee parking from visitor parking. If underground parking exists, attempt to accommodate visitor parking on the street.

11. Create as much offset from the street as possible. Overpressure is inversely proportional to the cube of the distance from the blast. Each additional amount of standoff provides progressively more protection.

12. Terrain, other buildings and barriers, and vegetation can provide some additional protection by absorbing and deflecting blast forces and debris. Be careful that vegetation does not provide a hiding place for a criminal or for a device.

Preparedness

1. Audit the physical security, incoming inspection, access, and internal controls of the facility. Analyze the firm's exposure to bombings.

2. Place bomb threat questionnaires and brief instructions at all security, switchboard, and reception stations.

3. Provide specific training to all security officers, switchboard operators, and receptionists who could receive bomb threats.

4. Establish who will evaluate threats. Train this person or an evaluation committee to assess the credibility of threats.

5. Decide what procedures will be followed when a threat is received and if a device is discovered, and what methods will be used to search.

6. Decide what conditions (if any) should trigger an immediate evacuation of the building, and list these conditions in a procedure for security or for an ERT commander. Companies may decide to identify certain elements as justifying evacuation of employees without approval from the threat assessment committee. Depending on the company's bomb-risk profile, these elements may include disclosure of the location of the bomb, the time of detonation, a motive, or an apparent familiarity of the facility.

7. Identify search methods and searchers before the incident. Train searchers on technique, bomb recognition, and safety. Most police departments will not search for you!

8. Identify and prioritize search areas before the incident.

9. Consider delaying a search to protect the safety of the searchers if the threat is credible and a short time limit is given.
10. Establish a procedure to track the progress of the search. Develop sectionalized maps and checklists.
11. Test all procedures.
12. Review plans, procedures, and contact phone numbers regularly.
13. Understand your community's capability to analyze postdetonation conditions and to monitor for other hazards, such as chemical agents, gases, radioactive materials, or secondary devices.

Response

1. Write down the exact time of a telephoned threat. Find out whether the call originated from within the facility or from the outside.
2. Record the caller's exact words. Permit the caller to say as much as possible, without interruption.
3. Ask the caller the following, and record the answers to these questions:
   - When will it explode?
   - Where is it located?
   - What does it look like?
   - Why was it placed?
   - Who is calling?
4. Attempt to transfer the call to the head of security or to a member of the threat assessment team.
5. Initiate a search for a device. Use a checklist to ensure that all prioritized areas are searched in the most expedient and thorough manner possible. If the caller tells you where the bomb is located, obviously check that area first.
6. Make other notifications as appropriate (to management, security, and the police).
7. If you find an object you suspect may be a bomb:
   - Evacuate people to at least 300 yards away and out of the line of sight or blast effect.
   - Dial 911.
   - Identify and evaluate the object.
   - Do not attempt to touch, move, dismantle, or pour water on any suspicious object.
   - If it is safe, open doors and windows around the area to reduce blast effect.
   - Isolate (secure) the area from entry.
   - Consider shutting down utilities and hazardous processes.
   - If it is safe, continue the search for additional bombs.
8. Relocate vital records, and back up computer systems if time permits.
9. Contact the public relations spokesperson or activate the crisis management plan.

10. Stage emergency response equipment and strategic resources.
11. Restrict access to the area.
12. If a bomb explodes, and people are injured, some experts suggest removing the victims immediately and securing the area, not treating them where they are found. Their rationale is based on the practice of some terrorists to place in the same area additional bombs set to explode later to kill and injure rescue workers and police.
13. If an RDD is suspected:
    - Move far away from the immediate area (upwind if possible). If this is not possible, go inside to reduce exposure to any radioactive airborne dust.
    - Cover or filter your mouth and nose to reduce the chance of breathing radioactive dust.
    - Once inside, remove clothing and seal it in a plastic bag (if you used material to cover your mouth, place this in the bag also). This practice can remove the great majority of exposure to dust and alpha radiation.
    - Shower to remove the remainder of the dust.
    - Building owners should shut all windows, outside doors, and dampers. Fans and heating, ventilation, and air conditioning (HVAC) should be turned off.
    - Monitor television and radio stations for news and instructions.

Recovery

1. Care for the injured.
2. Begin rescue operations only if properly trained and equipped, and if there are no other devices in the area.
3. Assess damage, including the structural integrity of the buildings.
4. Begin salvage and cleanup.
5. Test electronic and other sensitive equipment for blast damage.
6. Keep employees informed about the status of cleanup and future risk from any radiological residue.
7. Provide posttraumatic stress counseling for employees and rescue crews.
8. Begin relocation and reconstruction.
9. Investigate and prosecute every incident.

EARTHQUAKE

The release of energy caused by resistance to the continual shifting of large segments of the earth's crust (tectonic plates) is responsible for most of the world's earthquakes. The boundaries of these plates are fault zones. Although most major earthquakes occur near the fault zones, few places in the world are totally immune.

Major earthquakes can occur even many thousands of miles away from these boundaries; they are thought to be related to interplate crustal weakness. Such was the case for the largest earthquakes in the continental United States, along the New Madrid (Missouri) fault in 1811 and 1812. No portion of the United States or southern Canada is immune from the effects of earthquakes. The entire West Coast (including Canada, Alaska, Nevada, and Utah), the Midwest near the Mississippi River, and the East Coast north of Florida up to southeastern Canada and New England are susceptible to significant ground movement.

Damage from earthquakes is related to the amount of energy released, length of the fault rupture, depth and type of fault, velocity and acceleration, distance from the fault, soil types and conditions, and type of building construction. Various methods are used to measure earthquakes. The Richter scale indirectly measures the energy released from an earthquake by recording needle deflections on a seismograph. An earthquake with a Richter magnitude of 3 is barely felt unless you are very close to it. A magnitude 6 earthquake can cause major damage. The Richter scale uses a logarithmic progression. That is, the energy released from a magnitude 6 earthquake is not twice as big as that from a magnitude 3: it is about 100 times greater. The Great Sumatra earthquake of 2004 that generated the tsunami responsible for almost 300,000 deaths was estimated to be a magnitude 9.15, one of the largest since 1900. Some scientists placed the magnitude as high as 9.3. This was caused by a sea-bed rupture of nearly 800 miles with estimates of vertical movement in some locations of 50 feet, vibrating the planet by as much as ½ inch (1 centimeter). The shaking lasted for about 10 minutes. The earthquake released energy equivalent to 100 gigatons of TNT and caused a slight change in the earth's rotation.

The modified Mercalli scale is used to map areas of intensity, based on personal reports by victims and by inspection of the damage. It uses a scale from I through XII, based on the type of damage observed. Although not as popular as the Richter scale, it provides a better indication of actual damage.

Certain soil conditions amplify or attenuate seismic forces. Soil and bedrock in the eastern United States tend to transmit this energy over a wider area than that in the West, allowing an earthquake in Bolivia to be felt in Minneapolis. Sand and silt near unstable bay and river areas can cause liquefaction and soil failure. Liquefaction occurs when the ground motion causes sandy materials saturated with water to behave like a liquid.

Most modern structures are designed to withstand earthquakes. They may sustain heavy damage, but they should not collapse. The greatest danger is from objects such as parapets, signs, bricks, and glass falling off buildings. Older buildings, non-retrofitted buildings of concrete or tilt-up or non-reinforced masonry construction, can partially or completely collapse. The Uniform Building Code in the United States and the National Building Code in Canada contain maps that assign a seismic risk (based on expected ground movement, not probability) to various parts of the countries and define building codes to balance the known level of risk in these areas.

Earthquakes can cause a complete collapse of transportation systems: highways, overpasses, bridges, shipping ports, and airport runways can be damaged or destroyed. Infrastructure can fail—expect to be without water, sewer, and utilities for 3 days or more; police, fire, and hospital services can be destroyed or overloaded. Thousands can be made homeless or refuse to return to their homes, even if damage is minor. Dam and levee failures, train derailments, landslides, hazardous material releases and spills, uncontrolled fires, avalanches, people trapped in buildings, and people injured in falls could all result from an earthquake. The destruction generated by tsunami produced by underwater faults is well known.

Presently, forecasting is based on known fault locations and the historical recurrence rate of earthquakes on these faults. These predictions are difficult, in part, because many faults are not discovered until they cause the earth to move. A fault under Los Angeles, only identified in 1999, is expected to cause losses of up to $250 billion and result in between 3,000 and 18,000 fatalities.

Mitigation

1. Understand the seismic risk for your area and plan accordingly.
2. Identify structural and nonstructural hazards to facilities, occupants, equipment, and processes.
3. Ensure that automatic fire sprinkler systems are supported with earthquake sway bracing (see National Fire Protection Association [NFPA] Standard 13, Standard for the Installation of Sprinkler Systems). Pipes can break during an earthquake, causing water damage or inability to extinguish a fire.
4. Bolt bookcases; file cabinets; display racks; workstations; flammable liquid storage cabinets; wire or equipment racks; and other tall, heavy objects to the floor or wall. These objects can topple over during an earthquake, causing injuries to employees, blocking escape routes, damaging equipment, and delaying the cleanup and recovery process.
5. Strap sensitive, critical, or expensive equipment (including computers and servers) to desk, workbenches, or equipment racks. Place equipment that cannot be strapped or that must be moved often on seismic damping mats. Base-isolate (prevent the base from moving, or attach devices that allow the base to move independently) large equipment such as boilers, pumps, chillers, and backup electrical generators.
6. Identify and mitigate other nonstructural hazards, including light fixtures, storage racks, hanging objects, mirrors, and fireplaces.
7. Strap water heaters to the wall and install flexible connections to the gas line. Water heaters can topple over during an earthquake, causing fires and the loss of a potential source of drinking water.
8. Consider the installation of seismic switches that shut off hazardous process gasses or equipment when a preset acceleration is achieved.
9. Arrange for backup power generation.
10. Ensure you have a plan for redundant communications.
11. Prepare a mass casualty plan.

Preparedness

1. Stockpile supplies of food, water, lighting, and disaster first aid supplies. Employees may need to remain on site temporarily while transportation systems are being restored. Encourage employees to keep emergency supplies in their desk or car. At a minimum, this should include the following:
   - First aid supplies and prescription medications
   - High-energy, low-salt packaged, dried, or canned food

- Nonelectric can opener
- Portable radio with batteries
- Flashlight or light sticks
- Extra cash
- Tennis shoes
- Blanket
- Pocket knife
- Safety whistle (this should remain in the employee's pocket or purse)
- Picture of kids, family
- Out-of-town contact phone numbers
- Change of underwear and socks
- Toiletries and personal hygiene items
- Proof of residence, such as a water bill
- Drinking water—at least 1 gallon per person per day. Water supplies may not be available for 3 to 5 days after an earthquake

2. Encourage individual and family preparedness. Employees who know that their families and loved ones are safe at home are more willing to remain at work or to return to work soon. Conduct earthquake safety workshops and obtain information from the American Red Cross and the Federal Emergency Management Agency (FEMA) to distribute to employees.

3. Consult with a structural engineer to ensure that your buildings meet seismic safety standards. Arrange to have these engineers inspect the buildings after an earthquake.

4. Train personnel in the location of emergency shutoff valves for gas, electrical, water, and hazardous gas or chemical lines. Install flow valves on hazardous material lines to prevent uncontrolled discharge if the pipes are broken.

5. Train employees in first aid and cardiopulmonary resuscitation (CPR).

6. Maintain an adequate supply of cash.

7. Set up an agreement with the local authorities to accept a contracted structural engineer's judgment about the safety of your facilities after a major earthquake. Local inspectors may not check your building in a timely manner and, because of time constraints and workload, may be less likely to allow occupancy.

8. Consider purchasing earthquake insurance.

Response

If you are inside—DUCK, COVER, and HOLD:

- Take cover under a desk or table, or sit or stand against an inside wall (not inside a doorway).
- Hold tightly to the desk or table until the shaking stops.
- Move away from windows or objects that may fall on you.
- Do not run outside during the shaking.

- After the shaking has stopped, evacuate the building if structural damage is apparent.
- Avoid use of the telephone. Replace telephone handsets shaken off the hook.
- Do not use open flames.

If you are outside:

- Do not enter any building.
- Move clear of buildings, falling glass, utility poles, wires, and large trees.
- Get on the ground—DUCK, COVER, and HOLD.
- After the shaking stops, watch for falling glass, electrical wires, poles, and other debris.

If you are driving:

- Drive away from overpasses and underpasses.
- Stop in a safe place.
- Set the parking brake.
- Stay in the vehicle. If wires fall onto the vehicle, stay inside until rescued.

Recovery

1. Check for injuries.
2. Rescue victims, if trained and equipped to do so.
3. Check the structural integrity of the building.
4. Check for fires and other damage. Turn off gas only if a leak is detected.
5. Clean up hazardous material spills.
6. Turn off noncritical electrical equipment if power is out. When power is restored, turn equipment back on gradually.
7. Avoid the use of open flames. Do not use camp stoves or charcoal in an enclosed area, such as a tent.
8. Use caution when reentering damaged buildings; aftershocks can cause further damage.
9. Check that vents and exhaust pipes have not separated from water heaters or other machinery.
10. Listen to radio and television stations for official information.
11. Begin relocation and reconstruction if necessary. If employees are to occupy a partially damaged building, concentrate first on cosmetic repairs. No one wants to look at cracks in the wall during an aftershock.

CHEMICAL OR BIOLOGICAL ATTACK

Experts in national and international security affairs have long warned about the use of biological, chemical, and radiological weapons against our population. The ability to produce and use these methods of mass destruction is well within the grasp of terrorist organizations that often share a fanatical belief they are defending their religion and homeland. The present-day superterrorists are well educated, well funded, and better organized. Their intent is to cause violence within our cities and to destroy our businesses and way of life.

Authorities believe our greatest risk from these criminals comes from the deliberate destruction of one or more of the 850,000 facilities in the United States that deal with hazardous or extremely hazardous substances, causing the release of these materials into the community. Others point to the desire to use more exotic and more deadly substances such as nerve agents (tabun, soman, VX), vesicants (arsine, lewisite, phosgene), and biotoxins (Cryptosporidium, Ebola virus, smallpox).

Many authorities, however, believe that most of these groups lack the technical and tactical ability to use these nonconventional weapons to cause large numbers of fatalities. The Aum Shinrikyo, a Japanese religious cult, employed numerous scientists and spent $30 million in a failed attempt to cause mass casualties by the release of sarin in their subway system. One quart of nerve agent that contains roughly 1 million lethal doses can be brewed in a kitchen or garage, but more than a ton of nerve agent would be required to kill 10,000 people outdoors. City water supplies are considered safe because of the filtration, frequent testing, and dilution of any agents introduced into a reservoir, but agents and methods exist that may bring this safety into question. Most chemical and biological agents that present an inhalation hazard break down fairly readily when exposed to the sun, diluted with water, or dissipated in high winds. However, agents released inside a building will be less affected by these mitigating conditions.

The best defense a business can take to mitigate and respond to terrorist attacks is through improved security measures and business continuity planning. The purchase of gas masks by individuals is not recommended because they are often outdated and agent specific, and need to be fitted properly. Stockpiling antibiotics is also not recommended because people who self-medicate themselves or their children could do more harm than good.

Nevertheless, we should watch for the warning signs of an attack. A chemical or biological attack won't always be immediately apparent because many agents are odorless and colorless, and some cause no immediately noticeable symptoms. Hospitals and emergency response personnel may be the first to detect a problem. Sudden high levels of absenteeism may also indicate a problem. Indications of a chemical or biological attack include the following:

- Droplets of oily film on surfaces or water
- Unusual number of dead or dying animals in the area
- Unusual or unauthorized spraying (crop dusting) in the area such as over a city or stadium
- Victims displaying symptoms of nausea, difficulty breathing, convulsions, disorientation, or patterns of illness inconsistent with natural disease, such as blisters and rashes
- Mass casualties
- Low-lying clouds or fog unrelated to weather, clouds of dust, or suspended particles

- People unusually dressed (long-sleeved shirts or coats in the summertime) or wearing breathing protection in areas where large numbers of people congregate such as subways or stadiums
- Unexplained odors (smell of bitter almonds, peach kernels, newly mown hay, or strong garlic)
- Observation of dispersal devices or unexploded or unreleased material in its container

If an attack or exposure occurs, the best defense is to quickly evacuate the area to as far away from the agent as possible. If you cannot outrun a contamination plume, you may need to shelter-in-place. Protection of breathing airways and the skin is the next most important step a person can take in the event of a chemical or biological exposure. Professional response is similar to a major Hazmat incident.

Mitigation

1. Implement stringent security control of toxic and hazardous chemicals and biological products, especially if they are used or stored in large quantities, such as chlorine or ammonia, or if they are used to manufacture other harmful agents.
2. Store chemical tankers inside buildings or secured yards. Ideally, buildings should provide complete containment with blast-resistant design, and be equipped with a scrubber system sufficient to neutralize the entire contents of the tanker.
3. If possible, substitute less toxic substances for toxic chemicals in hazardous processes.
4. Minimize the amount of dangerous chemicals on hand.
5. Secure access to ground level air intakes. If possible, elevate or relocate them to the roof. HVAC systems can become entry points for chemical and biological (and also radiological) contaminants and distribute them throughout a building.
6. Install filters on air systems.
7. Protect building utilities from tampering, including water mains. Reroute pipes and other utilities underground and away from the exterior of buildings. Secure any access to these utilities.
8. Install HVAC venting and purging systems.
9. Design new construction with security and terrorism protection components in mind.
10. Isolate mailrooms from other parts of the building. If warranted, consider a separate air-handling system.

Preparedness

1. Understand the types of biological and chemical agents terrorists may acquire, including their properties, consequences, and response options.

2. Identify businesses in the immediate area that could affect your operations and employees if attacked.

3. Develop procedures to notify employees, the surrounding community, and response and regulatory agencies if the hazard can be created within your organization.

4. If at high risk, develop evacuation routes and procedures to move employees out of the area before exposure.

5. Assess the physical security of buildings.

6. Restrict access to mechanical rooms and building operations systems.

7. Identify safe rooms to use for sheltering-in-place. Develop procedures and training programs for employees.

8. Develop a means to quickly monitor conditions in your surrounding area. This could be assigning responsibility to monitor police and fire broadcasts, news channels, or other forms of rapid communications and information gathering.

9. Work with human resources or other departments within your organization to quickly identify periods of sudden, out-of-the-ordinary absenteeism among employees.

Response

1. Move upwind from the source of the attack.

2. If evacuation from the area is impossible, move indoors and upward to an interior room on a higher floor if possible (many agents are heavier than air and will tend to stay close to the ground).

3. Close building dampers, windows, and doors. Turn off HVAC systems.

4. Shelter-in-place. This means that once indoors, close all windows and exterior doors and shut off air conditioning and heating systems. Seal all cracks around windows and doorframes with duct tape. Seal all openings in windows and doors (including keyholes) with cotton wool or wet rags and duct tape. A water-soaked cloth can be used to seal gaps under doors if wide tape is not available. Cover bare arms and legs and make sure any cuts or abrasions are covered or bandaged. Choose a room with access to a bathroom and preferably with a telephone. If possible, store a 3-day supply of food, water, flashlights, radio, prescription medications, and other emergency supplies. In a business environment, close dampers and shut off HVAC systems.

5. If splashed with an agent, immediately wash it off using copious amounts of warm soapy water or a diluted 10:1 bleach solution. Taking off your clothing can remove roughly 80 percent of the contamination hazard. Place your clothing in a sealed plastic bag. If outdoors and you cannot go inside, attempt to locate a fountain, pool, or other source of water so that you can quickly and thoroughly rinse any skin that may have been exposed. If water is not available, talcum powder or flour are also excellent means of decontamination

of liquid agents. Sprinkle the powder liberally over the affected skin area, wait 30 seconds, and brush off thoroughly. If available, rubber gloves should be used when carrying out this procedure.

6. If in a car, shut off outside air intake vents and roll up windows if no gas has entered the vehicle.

7. In any case of suspected exposure to chemical or biological agents, medical assistance should be sought as soon as possible, even if no symptoms are immediately evident.

8. If anthrax is suspected:
   - Do not handle, shake, or empty a letter or package. Place it in plastic such as a zipper-lock bag or plastic sheet protector.
   - Wash your hands with soap and warm water for 30 to 60 seconds, and then wash your face. Blow and wipe your nose.
   - Close the doors and windows of the room where the package or letter is located and turn off air conditioning, heating, and fans.
   - Collect the names of all people who have had contact with the letter or package.
   - If you develop flu-like symptoms, see your health care provider immediately.

9. Call the police and fire department.

10. Monitor news broadcasts for additional information such as locations for vaccinations.

11. Maintain resource information for decontamination and the clean up of biohazards.

Recovery

1. A chemical, biological, or radiological attack can have a lasting effect. The effects of a biological agent or pandemic could persist for many months and spread around the world. Plan for an extended relocation.

2. Track employees who may be quarantined, killed, or evacuated. The temporary or permanent loss of personnel may become the biggest issue from a continuity standpoint. This will include the employees of vendors, suppliers, and customers.

3. Develop a mass casualty plan for the organization that considers the biohazard risk.

4. Establish work at home capability or alternate worksites (applicable to most all recovery planning situations).

5. Cross-train employees where possible.

6. In the case of widespread sickness, minimize direct contact with others. Set up video conferences, web casts, and so forth.

7. Disinfect phones and other work surfaces on a regular basis.

8. Maintain sufficient out-of-region resources to meet continuity objectives.

9. Keep copies of building design and engineering data off site and secured to aid in the decontamination of the building.
10. Once an affected building has been decontaminated, educate the users about the process and its safety to reoccupy.

EVACUATION PLANNING

Since the collapse of the Twin Towers, evacuation planning is no longer as simple as listening for the fire alarm and finding the nearest exit. Past practices, such as relocating two floors up and three below in a high-rise building equipped with a sprinkler system and pressurized stairwells, are coming into question. Even the admonishment to not use elevators for evacuation is under review (it is still not a good idea to use them in a fire evacuation). Greater emphasis will be placed on planning for the full and complete evacuation of high-rise buildings, despite the time this may require (estimated at about 10 seconds per floor to move down a stairwell, assuming everybody keeps moving).

Fire and smoke, as well as other hazards, can spread very rapidly, necessitating quick action to protect building occupants and move them away from the danger. However, the decision to evacuate a building should not be taken lightly; injuries can occur, productivity is lost, morale can suffer, and people may be less likely to treat a real evacuation seriously. Planners must analyze their hazards and decide how they need to structure evacuations for their environment and for each foreseeable situation.

Once evacuation plans are in place, employees must become aware of their responsibilities, i.e., how to recognize the need to evacuate, what is expected of them, where to go, and what to do when they get there. This is often accomplished by the publication of an Emergency Action Guide issued to new employees and reviewed during new hire training, exit signs, and markers indicating where evacuees are to report once outside the building. Visitors must also have this knowledge, usually accomplished by maps that contain brief instructions and evacuation routes posted inside entrances and elevator lobbies, instructions on the back of visitor badges, or brochures distributed during the visitor registration process. In high-rise buildings, color-coded symbols can be painted on the walls of stairwell entrances with instructions to proceed down the stairs until they find a like symbol. That floor becomes their evacuation relocation floor (examples include a green star, yellow triangle, red circle, blue square). The same symbol is located every three floors below, or the number of floors mandated by local regulations.

A primary responsibility after an evacuation is to verify that no one is left in the building or in the area vacated. In days past, this was done by taking a head count of those who were present at the assembly point. Although this may prove effective when dealing with small groups of employees, it can become impossible with larger and more mobile populations. In general, the better approach is to assign primary and alternate "floor wardens" to search their area of responsibility and to report to the incident commander (or equivalent) that no one is left in their section. It is their responsibility to ensure the safe and complete evacuation of all occupants in their assigned area. The search area must be manageable so that the warden is not in excessive danger from not immediately evacuating (obviously, the warden should be the last out). In high-rise buildings, additional wardens may be assigned to monitor stairwells to keep people moving, stationed at elevator lobbies, or assigned to help the disabled. Wardens should be clearly identified by wearing distinctive hard hats, colored vests, armbands, or other means of identification. Many large cities will mandate the number and type of wardens for buildings above a certain height, and may require additional duties. Floor wardens must be thoroughly trained in their responsibilities; in the emergency plan, first aid, CPR, and automated external defibrillator (AED) use; and in their own safety. They should understand the requirements of the ICS. Employees must be trained to acknowledge their existence and authority.

All evacuation planning should include visitors, vendors, contractors, and guests. This is especially true for the disabled; an increasing number of court cases are awarding damages to the disabled who were not adequately cared for during an evacuation.

Experience has shown that people tend to leave by the door from which they entered the building, even if they are just a few feet away from a more appropriate exit. Although people tend not to panic in emergency or disaster situations, evacuation planning and drills will reduce injury and confusion.

Mitigation

1. Ensure that the evacuation warning system and fire department communication panels are connected to backup power and that the systems will not be damaged in a fire or earthquake. Absent local regulations, panels should be located away from evacuation routes, noisy areas, or areas subject to blast damage.
2. Ensure that your evacuation planning conforms with codes and standards and with local regulations. The NFPA publishes the NFPA-101, Life Safety Code that contains many definitions, guidelines, and requirements for evacuation planning.
3. In high-rise buildings, assign a dedicated fire safety director whose sole function is life safety for the building occupants. This is required in certain jurisdictions.
4. To the greatest extent practical, upgrade life safety elements of the building. Strengthen evacuation stairwells, make them wider to accommodate a greater volume of people, improve lighting, and make them smokeproof.
5. In high-rise buildings (or single-story buildings for that matter), increase the evacuation discharge area to ensure it is large enough to accommodate a full evacuation.
6. Keep current on the latest standards and thinking about evacuation procedures.
7. Have backup (redundant) communication systems in place (public address system, elevator telephone system, and telephone system).
8. Ensure that stairwell doors unlock during an evacuation or fire emergency.
9. Maintain close ties with the fire department, and ensure that response plans are coordinated.

10. Determine whether part of the fire department's response plan includes a search of stairwells to ensure that all people have left the building.

Preparedness

1. Determine who will plan, coordinate, and authorize evacuations. Assign an alternate for this person. Employees should be trained to activate fire alarms if this method can be used to initiate an evacuation.

2. Develop a script to use when announcing an evacuation. For example: "Attention. All occupants of floors 10, 11, 12, 13, and 14 are to immediately enter the nearest stairwell and proceed down to the floor with the same evacuation symbol as the one for your current floor." Repeat the message several times.

3. Develop unobstructed emergency escape routes that comply with local regulations and that do not lead to or through hazardous areas. Ensure that your plan is consistent with legal requirements, such as for proper signage (illuminated exit signs), posted evacuation route maps, and number of exits. Never nail doors shut; clearly mark any doors and hallways that are not exits. If the emergency is minor, it may be necessary only to relocate employees to a safe location within the building. In high-rise buildings, it may be appropriate to evacuate only the three floors above the incident and the two floors below (some jurisdictions require a five-floor relocation: two floors above a fire floor and two below are moved down five floors).

4. Devise special procedures for employees to stay behind and shut down equipment or perform critical operations. However, delaying the evacuation of employees should be minimized, if not eliminated.

5. Select and train evacuation monitors or floor wardens and clearly define their duties. Floor wardens typically confirm that everyone in their assigned area evacuates, ensure that the escape route is safe, and assist where needed.

6. List methods used to notify employees of an evacuation. These can include activation of the fire alarm or other distinctive audiovisual warnings (be certain that warnings can be heard in all locations of the facility), public address system announcements (develop scripted warning messages that can be adapted to the immediate situation—see earlier), or word of mouth (use floor wardens or supervisors). Consider repeating the notification in a different language.

7. Identify employees with handicaps who may need assistance with evacuations. Assign "buddies" or floor wardens to assist these people. Purchase special equipment, such as evacuation chairs, to assist with their removal from the building. Train potential users of this equipment on its location and operation. Wheelchairs should not be taken into stairwells.

8. Establish assembly points. Show the assembly points in the Emergency Action Guide (if required) or on wall maps.[5] If feasible, place a sign, placard, or brightly painted markings on the ground to indicate where employees are to assemble. Assembly points should be a reasonable distance from the building, so that employees and guests are not injured by the emergency or other dangers. They should not be located where they could impede emergency operations or equipment.

9. Devise a method to account for employees and guests. This can be accomplished through a sign-in log, by having area supervisors taking roll calls of their employees and guests, or through the use of floor wardens.

10. Establish procedures and guidelines to follow when returning to the building. These should include a method to check whether the building is safe to reoccupy. Designate who has the authority to notify evacuees when it is safe to reenter.

11. Document your plan. Include the procedures in your Emergency Action Guide and list them on wall-mounted evacuation maps.

12. Conduct drills often, no less than annually. Local regulations may require drills more frequently. Drills should always be announced ahead of time, and no one should be excused from the drill. Drills should actually simulate an evacuation; that is, if evacuees are to relocate to a different floor, actually go to that floor.

Response

1. If it can be done safely, shut down hazardous processes before evacuation.

2. Train all employees to recognize the evacuation signal and to follow evacuation procedures. Include the following instructions:

   - All employees are to leave immediately by the nearest exit.
   - Do not return to the work area to obtain personal effects.
   - Do not use elevators.
   - Assist the disabled (if there are not floor wardens assigned to this task).
   - Go directly to the assembly point unless instructed otherwise (do not stop to talk with friends or to get something from a vehicle).
   - Do not reenter the building until the "all clear" is given. Your plan and instructions should designate who has the authority to issue the reentry order.

---

[5]Some states require employers to provide employees with written instructions to follow during an emergency.

3. Ensure that all employees and guests evacuate. Floor wardens will sweep an assigned area to check that all have evacuated and to help the disabled evacuate.
4. Notify emergency officials if anyone is not accounted for.
5. Keep employees informed until a decision is made to release employees to go home or to reoccupy the building.
6. Close stairwell doors after you enter; do not prop them open.
7. Remove high-heel or flip-flop shoes if descending stairs.

Recovery

1. Establish several methods to disseminate return-to-work instructions with employees if they are released to go home.
2. Review the plan and procedures at least annually or after any major change in the configuration of the building. Provide for the training of newly hired or transferred employees.
3. Consider instructing business continuity team members to take laptop computers with them when they evacuate as long as they don't need to return to the workstation to get the computer. Some organizations issue laptop computers to business continuity team members so that they can continue to operate at an alternate location. This is a controversial task because evacuees should not do anything that delays evacuation or carry anything with them that could cause injury or delay others.

FIRES

Fires account for the greatest number of losses to business, exceeding that from natural disasters. Sixty-four to 70 percent of businesses that have major fires never recover. They go out of business for good, primarily because of the loss of vital business records, particularly their accounts receivable files. Fire is also a major cause of accidental death in the United States. America's fire death rate is one of the highest per capita in the industrialized world, with just under 1.6 million fires causing the death of almost 4,000 and injuries to more than 18,000 people in 2003. This accounts for the third leading cause of death in the home. Injuries and burns from fires are often so hideous that their victims become reclusive; they do not want to be seen in public. The healing process from serious burns is said to be the most painful of any trauma. Direct property loss due to fires in the United States is estimated at $12.3 billion and at about $1.5 billion in Canada.

Many fires in business are caused by intentional or careless acts and by equipment failure, but bear in mind that arson is the preferred tool of many domestic terrorists, such as animal rights groups. Nasty labor disputes have resulted in and disgruntled employees have resorted to the burning of facilities. Recent, post 9/11 terrorist plots have revealed plans to drive gasoline tankers into the lobbies of buildings. Fires in public assembly areas and nightclubs have cost too many lives. Fires can accompany natural and manmade disasters, during times when the ability to control them is severely compromised.

Fires develop when a combustible fuel comes in contact with an ignition source. This does not need to be direct contact because radiant energy can be sufficient to bring a fuel up to its ignition temperature. Because we are surrounded by combustible materials, it is generally easier to prevent fires by controlling their ignition sources.

Mitigation

1. Understand your fire risk, the fire codes, and engineering to prevent the ignition and spread of fires. Retrofit as much as possible to bring buildings up to the latest fire codes.
2. Install fire extinguishers on every floor and near hazardous areas.
3. Conduct regular fire prevention inspections and immediately correct any hazards identified (see later).
4. Maintain all fire control devices and equipment in good working order and up to local codes and standards.
5. Establish a vegetation "clear zone" extending at least 100 feet from brush, fields, or forests. Use fire-resistant landscaping and building materials.
6. Ensure that the address of the business is clearly marked and that the fire department is aware of the locations of shutoff valves and hazardous materials. Work with the fire department ahead of time to develop "pre-fire" plans.
7. Establish no-smoking policies and permit smoking only in a fire-safe area.

Preparedness

1. Establish a written fire prevention plan. The plan should at a minimum include the following elements[6]:
   - A list of potential fire hazards and their proper handling and storage procedures, potential ignition sources (such as welding, smoking, and others) and control procedures, and the type of fire protection equipment or systems that can control a fire involving the hazards
   - Names or regular job titles of those responsible for maintenance of equipment and systems installed to prevent or control ignitions or fires
   - Names or regular job titles of those responsible for the control of accumulation of flammable or combustible waste materials
   - Housekeeping. The company must control accumulations of flammable and combustible waste materials and residues so that they do not contribute to a fire emergency. The housekeeping procedures shall be included in the written fire prevention plan.

---

[6]Adapted from the California Code of Regulations, Title 8.

- The company shall apprise employees of the fire hazards of the materials and processes to which they are exposed.
- The company shall review with each employee upon initial assignment those parts of the fire prevention plan which the employee must know to protect the employee in the event of an emergency. The written plan shall be kept in the workplace and made available for employee review.
- Maintenance. The company shall regularly and properly maintain, according to established procedures, equipment and systems installed in the workplace to prevent accidental ignition of combustible materials.

2. Install automatic fire detection, suppression, and warning systems (UL-listed monitored fire alarm systems with smoke detectors and sprinklers).
3. Establish guidelines and other written material for employees. Train employees in fire safety and evacuation.
4. Shut off utilities, processes, and electronic systems if evacuation is anticipated. Be aware that some utilities cannot simply be turned back on without inspection. Never delay evacuation if this cannot be done safely.
5. In the case of a wild land-fire, place escape vehicles in position before evacuation is ordered.
6. Ensure that electrical systems are properly maintained. Is lightning protection required?
7. Inspect gas lines, furnaces, and boilers for good condition and the absence of leaks; protect processes that use flammable materials from damage, both accidental and intentional.
8. Assign someone to ensure that soldering irons, hot plates, and other ignition sources are turned off at the end of the day. Security officers should be trained and instructed to check for fire hazards while on their rounds.
9. Control traffic and parking so that emergency vehicles have access to the facility.
10. Preregister with restoration service companies that specialize in the cleaning, dehumidification, and odor removal of electronic systems and recovery of paper documents. The components of smoke can severely damage electronic equipment even if not in contact with heat and flames.

Response

1. Obtain basic information about the location, type, and size of the fire if it is reported to you. Record the time, name, and extension number of the person reporting the fire. Ask if personnel have evacuated the area and if anyone is trapped or injured by the fire or smoke.
2. Evacuate the immediate area. If it can be done safely and quickly, open windows and close doors. Don't use elevators. If floor wardens are used to verify that everybody has evacuated, leave your office

door open so they can check that everybody has left. Stay low, near to the floor, if smoke and heat are strong.

3. Dial 911 or the number that directly accesses the fire department dispatch center. Remain on the line until the dispatcher hangs up. Some organizations will first send a member of management, security, or the ERT to evaluate the need to notify the fire department. This can be a dangerous and costly approach because fires can get out of control quickly.

4. Fight the fire, if you are qualified to do so. In many jurisdictions, the equipment and training required to fight a fire beyond the incipient stage (nonstructural, pre-flashover) is extensive.

5. Shut off heating, ventilation, and air conditioning systems if a potential for bringing smoke from the outside into the building exists.

6. Dispatch one or more people to meet the fire department and direct them to the location of the fire. The less obvious the route, the more personnel should be assigned to this task.

7. Remove any vehicles or other impediments to responding equipment.

8. If the building has a sprinkler system, assign a person to monitor the post indicator valve (PIV). This person should verify that the valve is functional and in the open position. Remain at this position to ensure that the valve is not prematurely shut.

9. Assign a person to check or start the fire pump, if one is utilized.

10. Determine whether equipment or hazardous processes should be shut down. Do so only if it is safe.

11. Move flammable or hazardous materials away from the area. Do so only if it is safe.

Recovery

1. Contact a restoration service that can repair smoke-damaged equipment, clean up water, and control mold and odor.

2. Prevent further damage to facilities and equipment from rain, theft, and other fire-related problems.

3. Restrict access to the damaged area.

4. Contact insurance carrier or broker.

5. Determine whether the fire will affect production schedules or the functions dependent on any equipment, materials, or product destroyed by the fire.

6. Ask the fire department if it or disaster recovery teams can remove critical equipment or records not damaged in the fire. Some companies place large red or yellow dots on this equipment so it can be quickly located and removed if access to the building is restricted.

7. Recreate any lost vital records.

8. Provide status reports to major customers and employees.

9. Quickly replace or replenish used fire control equipment or devices.

10. Begin relocation and reconstruction.

FLOODS AND HEAVY RAIN

Second only to fires, floods are the most common and widespread of all natural disasters in the United States. On a worldwide basis, floods account for 39 percent more loss of life than earthquakes. According to FEMA, flooding has caused the deaths of more than 10,000 people in the United States since 1900, with more than $1 billion in property damage each year. Flooding is a major component of almost 9 of every 10 presidential disaster declarations that result from natural phenomena.

The onset of a flood can be gradual or sudden, and it can occur in areas where it is not normally expected. Heavy rains are usually the root cause. A summer storm in Colorado in 1976 was responsible for the deaths of 100 campers and vacationers—water came rushing down a river without warning. Floods of this type can reach heights of 30 feet or more; they can result from the collapse of a dam formed by a landslide, ice, or a debris jam. Flooding can also result from the structural failure of permanent dams during an earthquake. High seas produced by major storms or storm surge can also cause flooding, especially in low-lying areas. Tropical storms can carry huge amounts of water.

Floods are usually easy to predict. Flash floods generally recur in the same location, but can happen within minutes of a storm. Where floods are prevalent, the National Oceanic and Atmospheric Administration (NOAA) maintains River Forecast Centers and provides reports on river levels, rainfall, and predicted weather. Up-to-the-minute information is also obtained from firsthand observations, monitoring police and emergency services radio communications, and river-level web sites. Some jurisdictions prone to flash flooding have installed horns and sirens to warn people of impending danger.

Area-wide flooding causes many more problems than simply getting things wet. The force of the water in a river at flood stage can move vehicles and destroy buildings and bridges by battering them with an incredible force of the flowing water, rolling boulders, or entire trees. It can isolate areas from fire and police response and sever communications lines. Floods can, and often do, spread disease and contaminate drinking water.

Despite all the water, fires can be a big problem during floods. Underground storage tanks containing hazardous or flammable liquids can be forced to the surface. Equipment can be damaged or destroyed by water and mud, which often are deposited everywhere. Finished product can be contaminated and vital records damaged or destroyed. The foundations of buildings can be undermined by the erosive effect of rapidly moving water. The growth of certain molds can be deadly, and the odor is difficult to remove.

Determine the flood risk for your site, for access to your site, and for any infrastructure that feeds your site. List the elevations of facilities, critical processes, power sources, or other items that may be affected by flooding and understand what actions must occur at differing water levels. FEMA, as part of the Flood Insurance Rate Map (FIRM) program, issues maps that show the flood hazard area of your community. These maps are used to establish your flood risk and are different from inundation maps that show the areas of potential flooding due to dam or reservoir breaks. Use caution, however, because the maps do not always consider the maximum flooding or indicate all areas where flooding has, or can, occur. The maps outline the areas subject to flooding based on 100-year and 500-year flood potentials. Other terms for a 100-year flood potential include a *base flood, baseline flood,* and a *1 percent chance flood.* The term base flood is increasingly used because it does not carry the misnomer that a 100-year flood is a rare event or that water levels will not be exceeded. A 100-year flood does not mean that this level of flooding occurs only every 100 years! The 100-year flood has a 26 percent overall chance of occurring every 30 years and can (and has) repeated itself in the same location within the same year. A 1 percent flood means that there is a 1 percent or greater probability of the mapped area being equaled or exceeded during any given year. A 500-year flood indicates a 0.2 percent chance of being equaled or exceeded in any given year.

FEMA's HAZUS software mentioned in Chapter 12 also contains a flood mapping component.

Mitigation

1. Maintain a flood mitigation and response plan and list the necessary resources (sandbags, pumps, dikes, etc.). It may be required before you can apply for federal aid. Coordinate your planning with local agencies.
2. Keep streams, culverts, and other waterways clean of debris or silt deposits that could restrict drainage or cause a creek to dam. Work with local governments to assist in the removal of vegetation or other impediments.
3. Install flood barriers, flood panels and shields, diversion dikes, or other architectural barricades. Flood-proof at-risk structures (water-seal walls, install flood doors, reinforce exposed walls). If located in an area prone to flash flooding, consider erecting diversion walls.
4. Don't develop or expand into flood-prone areas.
5. Ensure that power shut-down instructions are distributed to the appropriate personnel and that these people are trained in the shut-down authority and procedures.
6. Determine whether underground tanks need to be drained and refilled with water to prevent damage from flotation.
7. Develop a plan to evacuate animals or equipment if appropriate.
8. If critical materials and supplies need to be delivered to keep the site in operation, develop a contingency delivery plan.
9. Consider the purchase of flood insurance.
10. Build diversion channels; straighten river channels.
11. Increase sewer and storm-water drain capacity.

Preparedness

1. Know your flood risk and the risk for flooding to your location's access and escape routes.
2. Install monitoring devices or arrange to directly monitor local and upstream water levels. Monitor river levels at locations that may affect

your facility or site, at access points to your site, and at locations that may affect infrastructure. Monitoring should include physical and weather service or water resources predictions. Have plans in place with up-to-date phone numbers, radio frequencies, or other resources needed to scrutinize water levels.

3. Know the flood-warning signs and your community's alert signals.

4. Review the contents of your community's flood plan.

5. Preregister with restoration service companies that specialize in the cleaning, dehumidification, and odor removal of electronic systems and recovery of paper documents. Often, these services will provide a list of actions to take to prepare, collect, and preserve documents and equipment to prevent further damage. Write these instructions into your plan.

6. Be prepared for erosion or landslides.

7. Keep roof and parking lot drainage systems free of leaves and other debris or obstructions.

8. Inspect roofs for water stains, ponding, plant growth, or other signs of potential weakness.

9. Remove any yard storage from drainage or other low-lying areas. Coat exposed metal with grease if it cannot be moved to higher ground.

10. Anchor any items or equipment that may float away. (Large tanks of flammable or hazardous materials can and do float away.) Consider off-loading storage tanks of hazardous materials and separating chemicals that can react if mixed together. Also consider filling emptied tanks with water to add weight, if this will not contaminate, or cause an adverse reaction with, its contents when refilled.

11. Install check-valves in sewer traps to prevent flood water from backing up into the building.

12. Maintain an adequate supply of sandbags, if located in an area susceptible to flooding.

13. Maintain a supply of waterproofing materials to protect equipment in the case of water leaks.

14. Park vehicles on the escape side of bridges in case they are weakened by rushing water. Keep vehicles fueled.

15. Monitor conditions by AM/FM radio, television, or NOAA weather radio broadcasts (156.40 and 156.55 MHz are the main frequencies). Flood watches (flooding is possible) and flood warnings (flooding is imminent or occurring) are issued by the National Weather Service. Some amateur television clubs have rented helicopters to transmit live video of flood conditions.

16. Stockpile other materials in a secure, accessible location. These should include:
    - Cleanup and salvage equipment
    - Emergency food and water

- Flashlights and light sticks
- Portable water pumps and hoses
- Portable generator and fuel
- First aid supplies and snake-bite kits
- Shovels and tools (including chain saw and crowbars)
- Rubber boots and gloves

17. Inspect fire protection equipment. Test fire pump and ensure that it has sufficient fuel.
18. Sandbag protection equipment and backup generators if they are susceptible to flooding.
19. Move equipment and documents to upper levels or to higher ground. Preidentify the most important documents to move.
20. Back up data systems, and transfer operations to an alternate location.

Response

1. Immediately deenergize equipment if the flood is isolated to your facility because of sprinkler system activation, broken pipes, and the like. Cover equipment and product with waterproof sheeting.
2. Monitor conditions and escape routes.
3. Shut off electrical power and utilities if flooding is imminent.
4. Immediately evacuate to higher ground—flood waters often rise rapidly.
5. Watch for and avoid low-lying areas. Don't drive through flooded areas. If your car stalls, abandon it immediately. Six inches of rushing water can knock people off of their feet. Almost half of all flash flood fatalities involve the occupants of vehicles.
6. Don't attempt to cross flowing streams or to swim to safety.
7. Beware of snakes and other animals.

Recovery

1. Before entering a building, inspect foundations for cracks or other damage.
2. Assess damage.
3. Contact the restoration service provider. Ensure that facilities and equipment are cleaned, dehumidified, sanitized, and deodorized before allowing the reentry of employees and guests.
4. Do not turn on utilities until the structure, appliances, and utilities are dry and the building is checked for natural gas or propane leaks.

5. Be sure water supplies are safe to drink. Dispose of any food or consumables that may have been in contact with flood waters or mold.
6. Obtain permit and inspection procedures for repairs and reconstruction.
7. Retrofit structures during repair and reconstruction.
8. Search for properties that should be acquired in order to remove at-risk structures from the floodplain.
9. Begin mitigation planning to avoid a repeat of the same problems in future flooding.
10. If relocating temporarily, inform customers of the new location.
11. Inform employees when they can expect to return.

HAZARDOUS MATERIALS INCIDENTS

The response to a hazardous materials incident, typically a chemical spill or airborne release, can range from a relatively simple cleanup procedure to one that is complicated, dangerous, and requires extensive training and equipment depending on the type and amount of the substance. A *hazardous material* is simply defined is a substance capable of creating harm to people, property, and the environment. The Environmental Protection Agency (EPA) and the Resource Conservation and Recovery Act (RCRA) have more definitive definitions.

Companies that use hazardous materials or that generate hazardous waste are in some form regulated by a number of federal and local laws. Additionally, civil and criminal sanctions are available to authorities if the regulations are willfully violated. The mitigation and response procedures for companies that use large volumes of hazardous materials can go far beyond what we have listed here.

The following are basic procedures the firm can use to prepare and respond to chemical spills, toxic releases, and other hazardous materials emergencies. The reader should check with local regulations and match or exceed the level of response and training they require. Before attempting to clean up chemical spills, be sure to use the proper protective equipment. If there is doubt, call the fire department.

Mitigation

1. Construct dikes and secondary containment around hazardous materials storage areas.
2. Separate incompatible materials that may react if mixed.
3. Train staff in the proper storage, use, hazards, and cleanup of hazardous materials.
4. Minimize the amount of hazardous materials stored on site.
5. Find alternate chemicals that can be substituted for hazardous materials.
6. Store hazardous materials or materials in process inside some type of containment connected to a scrubbing system that will neutralize any leaks.

7. Ensure that hazardous materials and hazardous waste are properly marked, stored, and removed from the property in an expedient manner.
8. Equip chemical storage tanks and hazardous processes with seismic shutoff switches if located in earthquake-prone areas.
9. Build dykes or diversion around drains or implement some method to prevent hazardous materials from entering storm drains.

Preparedness

1. Identify the hazardous materials used on site and maintain material data safety sheets in strategic locations (EOC, spill carts, and the like). Include a list of their quantities and their locations. Be aware of hazardous materials located at nearby facilities or transportation systems.
2. Determine what federal, state, and local regulations the company must follow. Train staff to the highest level appropriate to the level of response and cleanup that may be required.
3. Work with local officials, such as the fire department, to coordinate preemergency plans.
4. Place detection equipment, spill kits, safety showers, and equipment at strategic locations.
5. Maintain the proper equipment for first aid and for the cleanup and decontamination of materials used in your operations.
6. Install panic alarms and closed-circuit television (CCTV) in hazardous areas.
7. Install wind socks and weather stations if appropriate.
8. Conduct joint training drills with city or county services.
9. Investigate the need to install a community warning system.
10. Inspect containers and piping for damage or leaks.

Response

1. Evacuate the immediate area (upwind if possible). If required, prepare for shelter-in-place (see later).
2. Isolate the area and deny access to any unauthorized personnel.
3. Decide whether outside assistance (fire department or Hazmat cleanup) is required.
4. Wear the highest level of protective equipment available or required.
5. Remove injured victims or personnel overcome by fumes if you can do so safely.
6. Apply first aid appropriate to the injuries.
7. Identify the materials, their properties and hazards, cleanup procedures, and toxicologies. Assume the substance is hazardous until you know

otherwise. Look for identification on containers (don't walk into the secured area), placards, or material data safety sheets. Smell is not always a reliable way to identify chemicals. Never taste an unidentified substance (during the early days of the semiconductor industry, a Ph.D. engineer could not speak for several weeks because of an acid burn on his tongue caused by an unsuccessful attempt to identify a spilled substance).

8. Eliminate ignition sources if a flammable liquid or explosive gas is involved. Stage fire control equipment at appropriate locations.

9. If it is safe to do so, mitigate or eliminate the source of the spill—that is, close valves, cap bottles, patch leaks.

10. Determine the size of the area affected and whether additional evacuations are warranted.

11. If properly trained and equipped, contain the spill. Do not let material go down a drain or into a waterway, basement, or confined space.

12. Use only properly trained, equipped, and certified personnel to clean up the hazard. Never work alone.

13. Make notifications to regulatory agencies as required.

14. Shelter-in-place procedures (see the previous section on Chemical or Biological Attack for additional measures):
    - Turn off HVAC systems.
    - Close all doors, windows, shades, blinds, and transoms.
    - Place wet towels in spaces under doors.
    - Block air vents with plastic, and place tape around doors and windows.
    - Cover your body with clothing as much as possible.
    - Avoid eating anything uncovered that may be contaminated.
    - Listen to informational broadcasts.

Recovery

1. Activate the crisis management plan if not done earlier.
2. Decontaminate responders, facilities, and equipment.
3. Repair damage.
4. File any required reports with regulatory agencies.
5. Monitor health and environmental problems caused by the incident.
6. Investigate the cause and take actions to prevent a recurrence.

HURRICANES

The combined energy of the atmosphere and ocean can turn a tropical storm into a massive center of power and destruction. These cyclonic storms, intense low-pressure centers with swirling arms of clouds and winds over 75 miles per hour, cause billions of dollars in damage to the coastal areas of the Gulf of Mexico, the Atlantic Ocean, and

Table 13.1   The Saffir-Simpson Hurricane Scale

Category	Winds	Description	Effects
1	74–95 mph	Minimal	No real damage to building structures. Damage primarily to unanchored mobile homes, shrubbery, and trees. Also, some coastal road flooding and minor pier damage. Storm surge 4–5 feet.
2	96–110 mph	Moderate	Some roofing material, door, and window damage to buildings. Considerable damage to vegetation, mobile homes, and piers. Coastal and low-lying escape routes flood 2–4 hours before arrival of center. Small craft in unprotected anchorages break moorings. Storm surge 6–8 feet.
3	111–130 mph	Extensive	Some structural damage to small residences and utility buildings with a minor amount of curtain wall failures. Mobile homes are destroyed. Flooding near the coast destroys smaller structures, with larger structures damaged by floating debris. Terrain continuously lower than 5 feet above mean sea level (ASL) may be flooded inland 8 miles or more. Storm surge 9–12 feet.
4	131–155 mph	Extreme	More extensive curtain wall failures, with some complete roof structure failure on small residences. Major erosion of beach. Major damage to lower floors of structures near the shore. Terrain continuously lower than 10 feet ASL may be flooded inland as far as 6 miles. Storm surge 13–18 feet.
5	>155 mph	Catastrophic	Complete roof failure on many residences and industrial buildings. Some complete building failures with small utility buildings blown over or away. Major damage to lower floors of all structures located less than 15 feet ASL and within 500 yards of the shoreline. Storm surge of 18 or more feet.

many other regions in the middle latitudes. Hurricanes are considered the most destructive type of weather condition because they are capable of spawning tornados in addition to high winds and flooding.

Although the center, or "eye," of these storms is clear and calm, the winds across the rest of their 50- to 500-mile diameters can range from 75 to 150 miles per hour or more. Hurricanes are classified according to their wind speed; however, the storm surge, moisture content, and other damaging factors are not precisely related to these classifications. The Saffir-Simpson hurricane scale (Table 13.1) is an example.

Despite the power of a hurricane, the storm itself usually progresses at a speed of only up to 30 miles per hour, at times stopping entirely. Although the strong winds account for many of the problems associated with hurricanes, high seas and flooding cause the most damage and loss of life. During storms of this magnitude, the level of the sea may rise many feet higher than normal (a phenomenon known as *storm surge*). Hurricanes can drop tremendous amounts of water. Hurricane Floyd in 1999 dropped more than 20 inches of rain in North Carolina, causing the worst flooding in their recorded history, after it was downgraded to a tropical storm. This flooding contributed to almost 50 deaths and more than $1.3 billion in damage.

An average of seven storms per season form during the months of June through November and last for a week or more. This life span allows forecasters to spot and track a storm and predict its landfall. The loss of life in hurricanes in the United States is decreasing.[7] More accurate forecasting and a likelihood of people to take the warnings seriously account for this improvement. A hurricane watch means that hurricane conditions may occur; a hurricane warning means that a storm is expected within 24 hours.

Mitigation

1. Develop equipment and manufacturing shutdown procedures.
2. Keep trees in good health and trim branches that may fall into the building or onto equipment.
3. Reduce the potential for wind-borne debris.
4. Clean drains and catch-basins.
5. Upgrade existing structures and ensure that new construction incorporates high-wind building standards. Provide information from FEMA and other sources that employees can use to strengthen the wind resistance of their homes.
6. Avoid construction in areas subject to the greatest damage.
7. Strengthen life lines (utilities, roadways, and vital equipment). Elevate equipment such as fuel tanks, fire pumps, and electrical generators. Enclose this equipment in some type of protection.
8. Set up alternate work sites and access to data systems.

---

[7]Hurricane Mitch, which devastated Central America on October 22, 1998, was a category 5 storm with 180 mph sustained winds. This hurricane reportedly killed more than 10,000 people in Central America and will serve for generations to show what hurricanes can do. "These deaths are not just numbers, they were children, moms, dads, and friends." Jerry Jarrell, Director, National Hurricane Center.

Preparedness

1. See the section on Floods and Heavy Rain for additional guidelines.
2. Check flood insurance policies for adequacy of coverage and exclusions.
3. Before the storm season, stockpile such extra supplies as food, water, battery-powered weather radio, cash, plywood and nails, ropes, sandbags, emergency lighting, and power. These items may be in short supply when a storm is forecast.
4. Determine where to relocate business operations and employees if ordered to evacuate.
5. Inspect the integrity of roof edging strips, drains, pipe racks, and sign and stack supports.
6. Cover sensitive equipment and finished goods with waterproof covers.
7. Train employees to prepack clothes, medications, pictures, cash, extra glasses, and baby supplies, in case they are required to evacuate from their homes.
8. Prequalify restoration and building contractors for the business and employees.

Response

1. Monitor the progress of the storm and estimate the amount of time required to perform all essential tasks.
2. If ordered to evacuate, do so immediately.
3. Move valuables out of the area or to upper floors if located in a flood or storm surge zone.
4. Secure or store furniture, planters, or other objects located outside the building.
5. Board up windows with plywood (taping does not work).
6. Turn off utilities and HVAC.
7. Arrange for extra security if necessary.
8. If required to remain, stay inside and away from windows, doors, and outside walls. (Employees should never remain alone. Do as much as possible to ensure their safety.)

Recovery

1. Wait until the storm has passed before beginning repairs and restoration.
2. Evaluate the structural integrity of the building and utilities.
3. Ensure that the site is safe for cleanup, salvage, and reoccupation.
4. Account for employees as they return from evacuation.
5. Begin relocation and reconstruction if necessary.
6. If vital records or other resources were moved to alternate locations, return these documents as appropriate.

SERIOUS INJURY OR ILLNESS

Injury and illness are probably the most common emergencies found in the workplace. Businesses lose billions of dollars annually in lost productivity, recruiting, and other ways because of injuries. The firm's response to an injury is very visible and if handled incorrectly may become an issue, producing poor employee morale, a labor dispute, or negative publicity, not to mention the well-being of the injured.

As we have often mentioned, during and after a disaster situation, businesses may not be able to rely on the availability of ambulances, paramedic response, or even fully staffed hospitals. These services can become quickly overloaded, responders and doctors themselves injured or killed, and medical facilities damaged. Even if services are available, their response may be delayed as a result of traffic congestion, damaged bridges and overpasses, lack of fuel, and other obstacles. All employers have a legal and moral responsibility to protect the safety and health of their employees. They must be prepared to deal with first aid issues, especially if located in areas subject to catastrophic disasters that go beyond the everyday cut finger. It is therefore important to train employees to a commensurate level in first aid and CPR. Because their prices are becoming more reasonable, businesses should also consider the purchase of an AED for use in cardiac emergencies.

Mitigation

1. Implement effective injury and illness prevention plans (safety programs).
2. Evaluate the risk for and type of injuries expected in your environment. Develop prevention programs, response procedures, and stock any special first aid supplies required to provide emergency treatment.
3. Purchase biohazard cleanup kits and train employees in biohazard management and in the use of the kits.
4. Ensure that the location address is prominently and clearly visible to the fire department and ambulance companies. If appropriate, know and provide Global Positioning System (GPS) coordinates to responding agencies if equipped to use this technology in the event that street markers can be destroyed.
5. Sponsor flu inoculation programs and health fairs.

Preparedness

1. If employees are exposed to hazardous chemicals, install eye-wash stations and emergency showers.
2. Inoculate employees at risk, if necessary.
3. Know any special conditions or needs of employees and make this information available to first responders or the ERT.

4. Train employees in first aid and CPR. Match or exceed the level of training required by regulations to compensate for any extended response times of emergency medical personnel.
5. Purchase and train employees in the use of AEDs.
6. Maintain first aid supplies at strategic locations. First aid and medical (hospital) services may not be readily available after a disaster. Most first aid kits are not designed for the type of injuries or delays in care possible after a disaster. At least one kit should contain disaster first aid supplies.
7. Control traffic and parking so that emergency vehicles have access to the facility.
8. Keep a supply of blankets and perhaps a stretcher or other transportation device.

Response

1. Contact, or request someone to contact, emergency services (fire department, ambulance) and report:
   - Your name
   - A description of the illness or injury
   - Your location
2. Stay on the line until the dispatcher hangs up.
3. Determine the extent of the injuries.
4. Provide first aid if qualified:
   - Control bleeding.
   - Check breathing.
   - Check circulation.
   - Treat for shock.
5. Remain with the victim:
   - Ensure that the victim is not moved (unless to protect from further hazard).
   - Obtain as much information about the victim's condition, history, and needs as required.
6. Keep those not involved in the emergency away from the area.
7. Send people to meet the fire department, paramedics, or ambulance at the driveway and front door of the building to escort them to the victim.
8. Assign someone to hold elevators at the ground floor.

Recovery

1. Follow biohazard procedures and regulations when cleaning up body fluids.
2. Notify regulatory agencies (such as the Occupational Safety and Health Administration [OSHA]), legal advisors, and insurance carriers if required.

3. Completely investigate the cause of all injuries and take steps to prevent their recurrence.
4. Retrain injured employees in safety procedures.
5. Activate the crisis management plan if injuries are major, if illness is widespread or controversial, or if fatalities are involved.
6. Provide post-traumatic stress counseling if necessary.

LIGHTNING

Lightning develops as a result of interactions in the atmosphere between charged particles, interactions that produce an intense electrical field (up to 100,000,000 volts) within a thunderstorm. These bolts produce so much heat (50,000° F) that the air explodes into rumbling shock waves. Lightning kills more people across the United States each year than tornados; it is second only to flooding in weather-related deaths. In the United States, most deaths occur in Florida, Missouri, Texas, New York, and Tennessee. Lightning is one of the leading causes of interruptions and other power-quality disturbances. The amount of lightning activity an area receives is measured in terms of *flash density*, the number of cloud-to-ground lightning strikes per square mile during a year.

Of all weather-related phenomena, lightning is associated with the greatest number of myths. Lightning can and often does strike in the same place twice, and it seeks the best conductor to the ground, not necessarily the tallest object. The tires of an automobile do not protect it from lightning; the metal body will conduct the charge away from its occupants—convertibles excepted, but why would someone drive with the top down in the rain? You can be struck by lightning from a storm many miles away, even if the sky overhead is clear. A software engineer in Silicon Valley, an area of California where lightning is uncommon, was injured when his computer exploded. The engineer was on the ground floor of a high-rise building that was struck by lightning.

In addition to injuries, lightning can cause fires, structural damage, and electrical power outages. Lightning caused a fire and explosion at the Jim Beam distillery that caused more than $1 million in damage and the loss of 800,000 gallons of bourbon. Included in the loss figure is the payment of a fine of almost $30,000 for the contamination of a nearby river when the bourbon was allowed to spill into it. Lightning can cause brick or concrete walls to explode.

Mitigation

1. Understand the frequency of severe thunderstorms in your geographical area and plan accordingly. If located in a high-risk area, have your facilities evaluated by engineers, and implement the appropriate mitigation.
2. Build structures that incorporate lightning protection if located in a high-risk area.
3. Consider the purchase and installation of a backup generator.
4. Upgrade fire protection systems.
5. Establish 24-hour guaranteed service or equipment replacement agreements with vendors.

Preparedness

1. Monitor the storm's conditions. The National Weather Service broadcasts continuous weather and warning information over the NOAA weather radio stations. Internet sites provide weather conditions and information, and local stations often broadcast severe weather information.
2. Install line conditioners that reduce the effects of power surges on sensitive and critical equipment, and connect critical equipment to uninterruptible power supplies.
3. Instruct employees to stay indoors. Outdoor activities should be discontinued immediately.
4. Avoid contact with pipes, railings, wire fences, telephones, other electrical equipment and appliances, and faucets and showers.
5. Train employees about lightning safety, to avoid open areas if outside, and to seek low ground in a wooded area (not under single tree). Employees should crouch close to the ground on the balls of their feet, but they should not lie down. Stay clear of metal objects; if in a group, move away from each other. Include first aid as a part of the training.
6. Maintain scheduled backups of computer, telecommunications, and other data systems.
7. Shut down systems if feasible.

Response

1. Bring employees inside the building; warn employees who may be about to leave that they should remain inside.
2. Contact emergency services if someone is struck by lightning.
3. Treat injured victims. Remove them from danger. Cardiopulmonary distress, broken bones, and burns are the most common lightning-related injuries. Look for entry and exit wounds.
4. Stand clear of windows, doors, and electrical appliances.
5. Unplug appliances well before a storm nears, never during.
6. Avoid contact with piping, including sinks, baths, and faucets.
7. Avoid the use of the telephone except for emergencies (lightning can strike telephone lines and subsequently injure people speaking on the phone, but this accounts for a small percentage of lightning injuries).
8. Check for and extinguish fires.
9. If power is lost, shut down equipment.

Recovery

1. Log all events, actions, decisions, and expenses.
2. Activate emergency response teams as required.

3. Assess and document damage.
4. Reestablish utilities.
5. Test and replace damaged equipment and connectivity. Restore data if any are lost.

TORNADOS

A tornado is a violently rotating column of air that makes contact with the ground. (A tornado that does not touch the ground is referred to as a *funnel cloud*; a tornado over water is a *waterspout*.) Tornados usually develop from severe thunderstorms. They originate in the right-rear quadrant of the storm cell or at the leading edge of a line of thunderstorms. Their paths are unpredictable, cutting straight lines of destruction, or zigzagging back and forth, or hopping and skipping around, even reversing general direction. They usually travel from the southwest to the northeast at an average speed of 30 miles per hour, but they have been known to remain stationary.

Although most tornados occur between 3:00 and 7:00 PM during the months of April, May, and June, they can occur any time of the day and anywhere in the world. The Texas Panhandle, Oklahoma, Kansas, Nebraska, Iowa, Missouri, parts of Arkansas, Illinois, and Indiana have the most tornados, although they form in every state, and at greater frequencies than previously thought. Southern Ontario accounts for one third of Canada's total tornados, with seven of the nine strongest occurring in this region.

Most (about 62 percent) fall into the "weak" category, with rotational wind speeds of 100 miles per hour or less. Only 1.5 percent of tornados are classified as "violent," with wind speeds reaching 300 miles per hour or more. These violent tornados account for almost 70 percent of the fatalities. It is the wind speed that accounts for most of the destruction, not the sudden drop in air pressure. Wind speed is greatest at the upper portions of the tornado. Flying debris is the cause of most injuries.

The Fujita scale (Table 13.2) is used to rate the intensity of a tornado by examining the damage it caused. It is interesting to note that the size of a tornado is not necessarily related to its intensity.

The National Weather Service warns against attempts to outrun a tornado in a vehicle because some tornados can travel nearly 70 miles per hour.

A tornado watch is issued when severe thunderstorms or tornados are most likely to occur and when conditions are favorable for their formation. A warning is issued when a tornado is detected on radar or reported by observers. Warnings and firsthand reports are received by the following means:

- NOAA Weather Radio broadcasts
- Community warning sirens
- The Internet
- Real-time Doppler radar
- CCTV
- Citizens band, amateur radio, and police communications frequencies
- Local television and commercial radio channels

Table 13.2   The Fujita Tornado Scale

F-Scale	Name	Wind Speed	Damage
F0	Gale tornado	40–72 mph	Some damage to chimneys; breaks branches off trees; pushes over shallow-rooted trees; damages sign boards
F1	Moderate	73–112 mph	The lower limit is the beginning of hurricane wind speed; peels surface off roofs; mobile homes pushed off foundations or overturned; moving autos pushed off the roads; attached garages possibly destroyed
F2	Significant	113–157 mph	Considerable damage. Roofs torn off frame houses; mobile homes demolished; boxcars pushed over; large trees snapped or uprooted; light object missiles generated
F3	Severe	158–206 mph	Roof and some walls torn off well-constructed houses; trains overturned; most trees in forest uprooted
F4	Devastating	207–260 mph	Well-constructed houses leveled; structures with weak foundations blown off some distance; cars thrown and large missiles generated
F5	Incredible	261–318 mph	Strong frame houses lifted off foundations and carried considerable distances to disintegrate; automobile sized missiles fly through the air in excess of 100 meters; trees debarked; steel re-enforced concrete structures badly damaged
F6	Inconceivable	319–379 mph	These winds are very unlikely. The small area of damage they might produce would probably not be recognizable along with the mess produced by F4 and F5 wind that would surround the F6 winds. Missiles, such as cars and refrigerators, would do serious secondary damage that could not be directly identified as F6 damage. If this level is ever achieved, evidence for it might only be found in some manner of ground swirl pattern, for it may never be identifiable through engineering studies.

The key to surviving a tornado is early warning and getting to an appropriate shelter.

Mitigation

1. Keep trees and shrubbery trimmed. Make trees more wind resistant by removing diseased or damaged limbs, then strategically remove branches so that wind can blow through.

2. Strengthen the wind resistance of buildings and consider the installation of shutters on windows.
3. Work with the local jurisdiction to establish tornado warning systems if none exist in the community.
4. Build safe rooms or tornado shelters. FEMA has publications that describe these steps.

Preparedness

1. Monitor weather broadcast stations and conditions. Obtain and use NOAA weather radios.
2. Learn what tornado conditions look and sound like (often described as the sound of a freight train or airplane). Tornados frequently emerge from near the hail-producing portion of the storm; wind may die down, and the air may become very still. An approaching cloud of debris can indicate the location of a tornado even if a funnel is not visible. A rotating protrusion of the cloud base may indicate a developing tornado, especially if the upper portion of the cloud is rotating. The cloud may turn a greenish color (a phenomenon caused by hail). Tornados generally occur near the trailing edge of a thunderstorm. It is not uncommon to see clear, sunlit skies behind a tornado.
3. Know the meaning of watches and warnings.
4. Identify a shelter room. A storm cellar or basement is best, but an interior room without windows on the lowest floor is the next-best alternative (excluding mobile homes).
5. Instruct employees in the meanings of local warning systems (sirens, a red bar on a television screen, and the like).
6. Construct storm shelters if none are close by.
7. Maintain supplies of food, water, and emergency lighting in case employees cannot leave immediately after. Power may be lost, so plan accordingly.
8. Keep plastic sheeting and other materials on hand to cover equipment or a damaged roof.

Response

1. Seek shelter in a storm cellar, basement, or interior room without windows. Get under something sturdy, such as a desk or table. If in a modular office, seek shelter inside a sturdy building if there is no storm shelter near by.
2. Occupants of a high-rise building should go to an interior room, hallway, or interior stairwell on a lower floor (if you have time). Center hallways are often structurally the most reinforced part of a building.
3. Stay clear of windows (do not open them), doors, and outside walls.

4. If outside, seek a safe place in a sturdy nearby building. As a last resort, take cover in a ditch or low-lying area; lie flat with your hands covering your head.
5. If in a vehicle, seek shelter in a ditch as described previously.
6. Leave mobile homes for a storm shelter or a safe place in a sturdy nearby building.

Recovery

1. Assess damage to facilities.
2. Begin search and rescue operations only if properly trained and equipped.
3. Treat injuries.
4. Avoid the use of open flames.
5. Monitor news broadcasts and the location of disaster assistance centers.
6. Form teams of volunteers to assist employees affected by the tornado.
7. Document damage for insurance reimbursement.
8. Clear debris.
9. Begin relocation and reconstruction.

WORKPLACE VIOLENCE AND CIVIL DISTURBANCE

Violence is the leading cause of death for women and the second-leading cause of death for men in the workplace. In 1994, robbery-homicide represented 9 percent of total crime but accounted for 75 percent of workplace homicide. With more than 2 million workers physically attacked, more than 6 million threatened, and more than 16 million workers harassed, it is not surprising that two thirds of workers do not feel safe at work.[8] The 1995 cost of workplace violence was $36 billion. Direct legal and medical costs totaled $250,000 per incident, and jury awards exceed $5 million. Firms that suffer an incident of workplace violence experience an 80 percent loss of productivity in the subsequent week. One San Francisco law firm went out of business because a former client, with no warning, using automatic weapons, killed and injured a number of attorneys and clients in what was regarded as a random act of violence.

Workplace violence is defined as violent acts, including physical assaults and threats of assault, directed toward people at work or on duty. It is classified by the relationship of the perpetrator to the workplace:

- Type I—no legitimate relationship to workplace, usually robbery or other criminal act (most common)
- Type II—customer or client, that is, a patient or passenger

---

[8]National Institute of Safety and Health, Bulletin 57 (Washington, D.C., June 1996); 1994 figures.

- Type III—employment-related involvement, that is, employee, ex-employee, domestic violence spillover

Mitigation

1. Observe good hiring practices, conducting criminal background verification and investigation where appropriate.
2. Establish policy that strictly forbids any form of workplace violence or possession of weapons and that requires employees to report threats and potential problems to management. Communicate this policy to all employees. Consistently enforce sanctions for noncompliance.
3. Foster a good working environment, with consistent discipline, a problem-solving rather than a blaming attitude, open communication, respect, and a nonauthoritarian management style.
4. Never terminate an employee in a manner that is degrading or does not allow him or her the full use of social services available.
5. Audit and improve physical security and access control.
6. Remove objects, such as decorative rocks, that can be used by demonstrators, activists, or rioters to break windows, injure employees, or damage other property.
7. Ensure good physical security, including proper lighting, good visibility, CCTV, and posted signs, as well as other protections such as minimizing cash, avoiding high-crime areas, and using multiple workers. Access control and panic (or robbery) alarms are helpful.
8. Consider the construction or designation of a hardened "safe room."
9. Isolate lobbies from the remainder of the building, offices, or facility.

Preparedness

1. Develop a comprehensive workplace violence prevention and management program.
2. Train supervisory and management personnel in the following areas:
   - Recognition of potentially aggressive behavior
   - Recognition of signs of domestic violence
   - Diffusion of aggressive behavior
   - Recognition of the importance of an immediate response to complaints
   - Other elements that may be required by OSHA, such as crime awareness, location of and operation of alarm systems, communications procedures, and late night procedures for type I violence.
3. Protect potential victims. Obtain corporate and personal restraining orders, provide special parking and escort privileges, change their

shift or work location, and consider providing the employees with personal protection devices.

4. Form a committee to evaluate and manage threats or acts of violence. The committee should include the employee's manager; human resources; and legal, security, and medical (employee assistance program representative, staff, or contract psychologist) personnel.

5. Train security officers to understand arrest policies, and to avoid the following:
   - Being provoked by name calling or derogatory remarks
   - Discussing or arguing the merits or issues of the dispute with protesters or picketers
   - Antagonizing demonstrators or picketers
   - Throwing back objects thrown at them
   - Attempting to take anything from a protester unless in self-defense
   - Making physical contact unless blocking a doorway (in this instance, they should not place their hands on a protester)

6. Develop intelligence programs if prone to civil unrest and demonstrations.

Response

Type I (Robbery)

1. Follow the robber's directions.
2. Do not argue or fight with the robber and offer no resistance whatsoever.
3. Speak to the robber in a cooperative tone. Let the robber know you intend to follow his or her instructions.
4. Never produce a weapon during the event.
5. Move slowly and explain each move to the robber before you make it. Avoid surprising the robber.
6. Do not follow or chase the robber.
7. Do not touch anything the robber has handled.
8. Close and lock all doors.
9. Call the police immediately.
10. Write down everything you remember about the suspect and the robbery.
11. Protect evidence.

Type III (Employment-Related Involvement)

1. Assess the situation (if safe to do so), but do not delay calling 911. In the case of a demonstration, determine the number of protesters, the purpose of the demonstration, and the affiliation of the groups involved.

2. If possible:
   - Isolate potential victims.
   - Lock all doors.
   - Close blinds.
   - Take cover.

3. Summon help:
   - Dial 911
   - Activate panic alarm if available.
   - Pull fire alarm (except in the case of protesters or if you don't want to place employees outside of the building). Some argue against this practice because it could cause stairwell doors to unlock.
4. Warn other employees.
5. Notify appropriate managers.
6. Account for employees and guests when able.
7. Do not interact with participants in a disturbance.

Recovery

1. Activate the crisis management plan.
2. Immediately clean up biohazards and physical damage.
3. Provide stress counseling for employees, victims, and families.
4. Review security and response procedures.

14

□□□
□□□
□□□

Business Continuity Planning

Eugene Tucker, CPP, CFE, CBCP
Contributing author

"A plan is nothing, planning is everything."
—Dwight Eisenhower

Business continuity planning is defined in many different ways, each reflecting its author's particular slant on contingency planning. Many of these definitions attempt to combine the definitions of continuity *planning* and of a continuity *plan*. There is an important distinction between the two.

Business continuity planning is a process that identifies the critical functions of an organization and that develops strategies to minimize the effects of an outage or loss of service provided by these functions. The most common strategies involve some type of third-party data center or alternate, off-site processing and alternate workspace to restore operations to a minimally acceptable level. In today's business environment, it is no longer acceptable to return to, or to achieve a minimum level of, service after a disaster. These companies wish to, or need to, maintain operations at the current level or to take advantage of the disaster by the existence of the plan to gain market share over the competition. *Disaster recovery planning* is really synonymous with business continuity planning, but the term is a product of the data center. It represents the idea that recovery planning is important only to telecommunications and data centers. Business continuity planning implies recovery planning for all the critical functions or business units of an organization. Today, these terms are increasingly drifting apart. *Disaster recovery* refers to the reestablishment or continuity of information technology and data systems; *business continuity* refers to the recovery or continuity of business unit operations (systems versus people).

A *business continuity plan* is a comprehensive statement of consistent action taken before, during, and after a disaster or outage. The plan is designed for a worst-case scenario but should be flexible enough to address the more common, localized emergencies, such as power outages, server crashes, and fires. Although the actions listed in the plan contain sufficient detail to implement strategies designed to recover critical functions, they are more guides than inflexible dictates. Because it is not practical to plan for every type of contingency, and because each disaster has its own set of conditions, the ability to modify the plan must be incorporated.

179

Although a recovery plan is important, it is the planning process that returns the greatest value. This distinction is often missed by both planners and end users of continuity plans. The identification of critical functions, the thought and analysis behind the development of the strategies designed to recover the functions, and the knowledge of why one particular strategy was selected over another are not always apparent from simply reading the plan. This is valuable knowledge when last-minute decisions are required to adapt the plan to a particular situation. The planning process is also a training exercise. The participants must think through contingencies, so that the actions required to recover from them will be already familiar. Reading the plan for the first or second time just after the disaster will provide for a less than effective recovery. This is assuming, of course, that the plan is not buried under a hundred tons of rubble.

WHY PLAN?

Responsibility for contingency planning often resides with the risk manager, the chief financial officer (CFO), or the data center manager. Security managers are, however, increasingly taking the role of plan developers. Their experience with the protection of assets, involvement in the identification and the mitigation of risk, and emergency response duties makes them logical choices for this role. The ability to work effectively with all levels of management is a required trait for security managers, a trait that all successful ones possess.

Some types of businesses, such as financial institutions and industries regulated by toxics laws, are required to maintain continuity plans. Businesses are increasingly regulated by laws and standards, many differing widely in their approach and requirements. Some are intended to be industry specific and others broad based. Some use differing terminology, or try to package the same methodologies in different looking boxes.

In any case, even in the absence of regulatory requirements, it makes good business sense to maintain a continuity plan. The cost of downtime, the cost of reconstructing lost data, and the loss of cash flow can severely damage many organizations, even beyond their ability to recover. If they are unable to operate, retail and transportation operations can lose an average of more than $100,000 per hour, high-technology manufacturing $200,000 per hour, and financial brokerages more than $5 million per hour. The costs of rebuilding 20 megabytes of data are shown in Table 14.1.[1]

Without continuity planning, the organization may lose its competitive advantage, valuable employees, and future research. Organizations cannot insure against lost customers or a diminished public (customer) image. History consistently shows that between 35 and 50 percent of businesses never recover after major disasters.[2]

---

[1]*NSCA News* 1990 figures adjusted for 1995 inflation, from a presentation by Roger N. Farnsworth, Business Recovery Managers Association, 1996. Note: 20 MB of information is not a large amount of data.

[2]These figures are often reported in the industry and vary widely. There is some discussion about their validity. In any event, almost every major disaster has more than one example of a business that did not recover.

Table 14.1 Costs of Rebuilding 20 Megabytes of Data

Function	Time Required	Cost
Sales/Marketing	19 days	$20,000
Accounting	21 days	$22,000
Engineering	42 days	$114,000

Other rationales for continuity planning include the following:

- Fulfill requirement by financial auditors or by customers
- Prevent the loss of market share
- Capitalize on the lack of planning by the competition
- Uphold fiscal responsibility
- Maintain stockholder liability
- Fulfill regulatory requirement
- Retain key employees
- Prevent the loss of research
- Help ensure the safety of employees
- Preserve customer confidence
- Assist in the overall economic recovery of the community
- Assist in a quick and orderly recovery after a disaster
- Minimize the economic loss (devaluation) to the firm

THE PLANNING PROCESS

The basic steps involved in business continuity planning are simple, although their implementation can be complex and time consuming. The critical functions of the organization are identified and ranked according to their value to the organization or to their interdependencies with other critical components. Cost-effective strategies for recovering the critical functions to an acceptable level are evaluated. Once the recovery strategies are chosen, a plan is developed to implement the strategies. The plan is tested (the proper term is *exercised* or *simulated*), and provision for maintenance of the plan is established.

Before these steps commence, it is important to identify physical or procedural hazards that could cause an outage or delay the recovery process. When dealing with multiple sites, the planner should visit each location and conduct an inspection for these hazards. This inspection should identify single points of failure in critical systems, and it should produce a set of recommendations to mitigate the results of the hazards identified in the business impact or risk analysis. This is often included as part of the business impact analysis (BIA).

Next, the organization must prepare to respond to the disaster or to the emergency when it happens. The goals of emergency response are to protect the health and safety of employees, guests, and the community and to minimize damage to the organization by stabilizing the situation as quickly as possible. Response planning is not continuity planning, but the two plans can be integrated.

Once the disaster or emergency is stabilized, recovery and restoration will begin. The terms recovery, resumption, and restoration refer to separate phases of the organization's return to predisaster service levels (although some planners use them interchangeably). *Resumption* embraces the initial, short-term strategies and steps to get back into production as quickly as possible. Moving to a hot site (a separate building or office area with duplicate, or equivalent, equipment already installed, waiting for emergency use) and transferring production to a satellite facility are examples. *Recovery* and *restoration* refer to the long-term strategies and steps the company will follow to reestablish its normal goals, service, or production levels. The replacement of a production line, installation and testing of replacement equipment, and the construction of new facilities are examples. In this text, we will use these terms interchangeably.

PROJECT MANAGEMENT

Business continuity planning projects, if not properly managed, will lose momentum, languish, and die, or assume such a negative tone that the participants become hesitant to complete the project. Information and the strategic mission of an organization can rapidly change, so that once the project is started, any significant delay will cause the end product—a business continuity plan—to be outdated before it is completed.

Project management is a major skill, and it is required of anyone who undertakes responsibility for business continuity planning. It is a partnership between members of management, outside services and vendors, employees, and sometimes regulatory agencies. The ability to schedule and manage resources, time, and people will help bring the project to a successful conclusion.[3]

In broad terms, the following steps are followed to produce an effective plan[4]:

1. Identify the planning coordinator.
2. Obtain management support and resources.
3. Define the scope and planning methodology.
4. Conduct risk identification and mitigation inspections.
5. Conduct a business impact analysis (BIA).
6. Identify critical functions.
7. Develop recovery strategies.
8. Set up recovery teams.
9. Develop team recovery instructions.
10. Collect resource information.
11. Document the plan.
12. Train recovery teams.
13. Exercise the plan.
14. Maintain the plan.

---

[3]Business continuity planning is an ongoing process of maintaining resource information and reexamining the basic strategies used to develop the plan. The plans must be "living" documents; the project is never "concluded."

[4]The Disaster Recovery Institute International (DRII), in an attempt to standardize planning methodologies, has issued a common body of knowledge, which may list steps different from those listed here. See Appendix C, Professional Practices for Business Continuity Planners. Regulations and other standards may also dictate a different approach.

1. Identify the Planning Coordinator

A person within the organization is designated as the planning manager, coordinator, leader, or other appropriate title. This person is responsible for the management of the project (that is, the completion of the plan) and possibly for coordinating or leading the recovery effort subsequent to a disaster. The coordinator may also have major responsibilities for plan activation. Ideally, this person should be a management-level employee who has good people and project management skills and a good understanding of the organization, and is detail oriented. A vast technical knowledge of risk and computer and telecommunications systems is not necessary, but it can be helpful. If a planning committee or a steering committee is established, this person should lead the committee. In a company with a dedicated business continuity planning department, this person is most likely the department manager, but individual planning coordinators may be assigned on a regional or national basis.

2. Obtain Management Support and Resources

No planning effort or project will be successful without the support of upper management. This support must be communicated to all levels of management. Most agree that the development of a business continuity plan is a noble project, but all too often other priorities take precedence if participants are not held accountable to time lines and milestones. Its timely completion should be included in the goals and objectives for all expected participants. Although the best results are obtained by motivating participants in a positive manner to complete their development tasks, it always helps to carry the big stick of upper management support behind your back. Absent the support of the board of directors or senior (executive) management as a group, it is often necessary to get the attention of and support from at least one member such as the CFO. This person becomes the business and continuity plan "sponsor."

In situations in which the driving force behind the project is not senior management, the results of a BIA should demonstrate to senior management the potential impact to the organization if all or a portion of the company could not operate. Their support is often obtained as a result of this demonstration. Before conducting the BIA, you may need to define the scope of the project.

3. Define the Scope and Planning Methodology

It can be a daunting, if not impossible, task to produce plans for a large, worldwide corporate structure unless the job is accomplished in small pieces. Narrow the scope of the project to a single division, site, or building, something small enough to allow a positive outcome. A successful project will add momentum for the completion of subsequent projects throughout the remainder of the organization.

The selection of a starting point may be dependent purely on need or on the degree of risk. The risk or BIA will show where the planner's energy is best directed. Be prepared to go beyond the established scope when identifying interdependencies and strategies.

The scope of the project may be to develop continuity plans for a manufacturing site in California, but the best strategies may involve the transfer of administrative functions and financial computing to the company's site in Illinois, and the transfer of manufacturing to Nevada.

Many methods exist to manage the project. These include the following:

- Planning standards for the business units or divisions
- Steering or planning committee
- Facilitation of internal development
- Use of template plans

In large organizations, corporate continuity planning policies and procedures are developed and instituted. Planning standards can be successfully developed even when the scope of the project involves many locations or business units. People responsible for planning at each location or division are chosen. Planning guidelines, methods, and templates are developed and distributed. The corporate planner then trains these people in the expectations of the project, monitors their progress, and helps to implement the plans. When this method is used, plans for the entire organization are developed quickly (usually a 2-year process). Greatest success is achieved when the corporate planner develops the "basic plan" (see Chapter 16, Plan Documentation) and allows the divisions to concentrate on planning for only themselves.

Many businesses form a planning or steering committee composed of the company's or division's top managers to discuss business impacts and select strategies, and they assign the documentation of the plan to individual functional mangers. The business continuity planner will coordinate the efforts of the committee, business units, and outside consultants. A variation of this method is often applied at the business-unit level, where a committee is formed to identify critical functions and strategies. Steering committees work best when the organization or planner has little knowledge of continuity planning principles and strategies or cannot gain an understanding of the operation or needs of the business unit.

The use of planning committees is suggested in some standards and regulations. Many planners insist that committees are the best way to produce effective continuity plans. Fundamental problems, however, can arise with the committee approach. Committees can delay decisions, and corporate policies can dilute the effectiveness of the process. Often, much time is wasted discussing administrative issues.

Another planning method places the planner in the role of a facilitator. The planner must bring his or her project and people management skills together to assist each individual business unit to develop its own plans. Each unit or recovery team must "own" its portion of the plan and become familiar with its contents. Working as a facilitator, the planner can tap the individual managers' or business unit leaders' technical knowledge and experience to guide the creation of their plans, while keeping them consistent in direction and format with the overall planning structure.

Many business continuity planning consultants promise to come in and develop a plan for your organization within 3 months, or sell you a template plan in which you just fill in the blanks, but your organization will gain very little from their product. The individuals responsible for its implementation will have little familiarity with, or commitment to, its contents. It becomes the consultant's plan, not your plan.

Consultants should act as facilitators and as advisors about the most effective strategies, if they are to deliver effective plans.

Use great caution when using templates (for instance, external or internal plans from other organizations). They are meaningless if not adapted to local conditions and needs. They do not require the "thought process" necessary to provide the training needed to execute the plan effectively after a disaster.

Once the scope and methodology are determined, select the managers who will participate in the process. This may be the same group that participated in the BIA. In most cases, they will be the functional or department heads. Schedule interviews of at least 1 hour's duration to discuss the following:

- Business impacts
- Critical functions
- Recovery strategies
- Expectations and needs of the manager
- Description of the project
- Questionnaires or resource forms
- How the function operates under normal conditions and how the product or service is delivered or produced
- Interdependencies—for instance, where the input comes from and who gets the function's output. (For example, a manufacturing line might get its input from raw materials stores. Its output goes to Quality Assurance. If raw materials stores are not recovered, or are not recovered first, manufacturing may have difficulty with its recovery. If a particular software application stops processing, other applications that depend on its output may stop or return inaccurate results.)

Schedule meetings over a 2- or 3-day period with participants from the following departments:

- Information Technology (IT) or the local area network (LAN) administrator
- Human resources
- Legal
- Facilities
- Management or site manager
- Manufacturing and operations
- Finance (accounts receivable and payable, cash management, and payroll)
- Telecommunications
- Risk and insurance manager
- Security and safety
- Other pertinent functions

Before the meetings, issue a letter or memo to the participants to introduce yourself and let them know what you expect. Briefly outline the process so they can begin to think about their responses.

4. Conduct Risk Identification and Mitigation Inspections

The more hazards and risks you can identify and mitigate beforehand the more you will minimize the effects of the disaster, allowing for a faster recovery.

Inspect the buildings, grounds, and community for any hazard that may injure employees, damage equipment or facilities, or cut off the supply of materials, resources, or services. When searching for these hazards, the techniques learned from scenario planning are useful. Think through the causes and effects of likely scenarios and offer recommendations to mitigate their effects. For example, some of the typical effects following a major earthquake may include the following:

- Structural damage and displacement
- Post-traumatic stress
- Loss of utilities (gas, water, electrical power)
- Disruption of communications
- Transportation difficulties
- Inflated prices for goods and services (in a cash economy)
- Human resource problems
- Overloaded and nonresponsive governmental services
- Victims trapped under structures and debris
- Fatalities, shortage of hospital beds and medical assistance
- Disruption of routines
- Difficulty obtaining food and water
- Uncontrolled fires
- Increased illnesses
- Damaged or destroyed product and raw materials
- Canceled orders
- Loss of vital records

A review of the literature on the effects of a flood would make clear the need to control mold, mildew, and snake bites, but such benefits of experience may not be available in print on the results of a failure of a proprietary process or a hazardous substance spill. Scenario planning will help you foresee the post-disaster conditions that must be considered.

Question the general manager, facilities manager, and other appropriate people on what could prevent emergencies, outages, and disasters. Does the company have an evacuation procedure? Are first aid, food, and water supplies stockpiled? Are critical systems or equipment connected to uninterruptible power supplies? Are computer files backed up on a regular basis and stored off site?

5. Conduct a Business Impact Analysis

When relevant risks and hazards have been identified, submit a report to the steering committee, senior management, or the sponsor of the project outlining recommendations to mitigate the hazards. This report can be combined with the results of the BIA, especially if the analysis has been completed informally, as is too often the case. The BIA identifies the *recovery time objectives* (RTOs) and is used to identify critical functions. See Chapter 15, Business Impact Analysis. Step number 4 above is also often included as part of the BIA.

6. Identify Critical Functions

The identification of critical functions is a major result of the BIA. Many planners believe it is a waste of time, effort, and resources to include in the plan functions that are not critical to the organization. Equipment and space at a hot site or other alternative location are expensive and limited; therefore, priority is given to the most important functions and employees. Remember that recovery operations are time sensitive. In many cases, there must be a logical sequence (order) of recovery actions, especially on the information technology side. Others argue that if a function is not critical, it should not be a part of the organization in the first place. I believe that every function should have a plan, but not necessarily a seat at the alternate site.

Generally speaking, a critical function can be a process, service, equipment, or duty that would have one of the following impacts on the company if the function were lost or if access to it were denied:

- Affect the financial position of the company
- Have a regulatory impact
- Reduce or destroy public or customer image or confidence or sales

7. Develop Recovery Strategies

Selecting the best continuity strategy, or group of strategies, is also key to an effective recovery. This is the heart of the continuity planning process. Despite their cost, continuity strategies are often chosen based on desire, what others do, or what seems expedient at the time. The method you use to continue the operation of critical functions must be based on cost versus benefit, technical feasibility, the results of the BIA, RTOs, and the strategic vision of the organization. This is often mixed with the function or process owner's personal resources (relationships with vendors, other business managers, property owners, and educational facilities). The strategy must be realistic and adhere to any assumptions contained in the plan. Strategies should not require employees to radically change their normal work habits or routines. They should not require extensive training subsequent to the disaster. During the BIA interviews with managers, discuss the alternatives they believe can be used to continue their operations for the short term and what actions are necessary to reestablish full operations at the present, alternate, or a new location.

Most large organizations use two high-level, organization-wide (or region-wide) continuity strategies; one for information technology (computer systems, software, data infrastructure) and one for alternate work space. These usually involve alternate processing capability and recovery space at a location outside the area subject to the effects of potential disasters (this is where an understanding of the effects of individual hazards becomes important). After a disaster, information technology operations are transferred to an alternate processing site, and selected recovery team members go to the alternate work space location, connect their computers to the alternate processing site, and continue operations until the damaged location is repaired or replaced.

Individual functions, business units, or data processes can have their own strategies in place that may differ from or complement the organization's overall plan. The organization may plan to relocate team members to an out-of-state convention center, but the technical support group may simply temporarily transfer their operations to a third-party call center or have other technical support groups in the company located in other regions take up the slack, or send the disaster-affected technical support employees to these locations.

The BIA may indicate that certain business processes cannot be out of service for more than 1 hour (or less) with very little data loss (recovery point objective), but other computer applications can disappear for 1 or 2 days without serious consequence (I will call these applications *Tier 3*). In this case, a *high-availability solution* is devised for the processes with low outage tolerance (high RTO); for example, the construction of a second data center at a remote location that receives copies of data as they are processed is on stand-by in case of a disaster. If there is a disaster, the second data center becomes the primary. This helps ensure a minimum of data loss. Copies of the Tier 3 applications are stored at the second data center and loaded on the system. Backup tapes from the Tier 3 applications are obtained and shipped to the second data center (if not stored there) and reloaded within its RTO. Processes are assigned to a tier, and different recovery solutions are devised for the groups to align costs with the need. Data from a Tier 2 application could be sent to the alternate data center by a batch method, wherein the backup data are stored until transmitted all at once, at certain time intervals.

High-availability solutions are generally the most expensive. A short RTO and recovery point objective (RPO), or lower outage tolerance, will result in a higher cost of the continuity strategy. Second data centers, as mentioned previously, are generally the best solution if no data loss is an issue, but are usually the most expensive. Third-party hot sites, network storage solutions, load balancing or server clustering (connecting multiple servers that share the processing and storage over a distance), and other information technology strategies are available. If the planner does not have the technical knowledge or if the expertise is not available internally, it is best at this point to bring in consultants.

During the interviews with managers, discuss options they believe can reestablish temporary (short-term) operations. Also discuss how the managers expect to implement the options and how long it will take. Repeat this process for a long-term outage. Very often the function leader has been through some type of outage or knows someone who has. What did they do to reestablish operations? Discuss the feasibility of other strategies, and pick the ones that will work best, based on recovery needs and requirements (for instance, RTO).

Complete a cost/benefit analysis for each strategy, using information from the BIA. If the loss of the function will cost the company $100,000 after 10 days, it makes little sense to spend $300,000 on a strategy that will put it back in operation in 1 day, especially if less expensive strategies exist to resume at least partial operations in, say, 8 days. Once the analysis is completed, select the best recovery strategy.

It is important, of course, that a strategy can be reliably implemented. If the recovery strategy for a West Coast technical support center is to transfer support calls to a center on the East Coast, be sure of the following:

- Its telephone equipment has the capacity to handle the extra volume of calls.
- The East Coast support staff is knowledgeable about the products supported by the West Coast center.
- Provisions are made for the difference in time zones.
- Extra staff is available on the East Coast.
- Manuals and documents are available.

The following list represents a small number of strategies the planner can select to recover critical functions, data, and equipment. It is by no means exhaustive, and each entry should be researched by the planner to determine whether it is the best strategy suited to the situation. During the hazard inspection or BIA, determine to what extent any of these strategies or redundancies are already implemented:

- Hot, cold, and warm sites
- Relocation
- Work at home
- Telecommunications
- Third-party manufacturing
- Purchase of material from competitors
- Data systems
- Revert to manual methods
- Virtual manufacturing
- Workforce management
- Reciprocal agreements
- Equipment rental
- Rescheduling production
- Reallocation of resources
- Service-level or quick-ship agreements

Hot, Cold, and Warm Sites

A *hot site* is an alternative recovery location prepared ahead of time, in this case with computers, servers, or a mainframe, and related equipment such as hardware and telecommunications. Hot-site vendors exist to provide this service on a first-come, first-served basis. The hot sites typically include a limited number of workstations and both data and voice communications infrastructure, enabling the organization to relocate employees temporarily. Organizations pay a subscription fee to the vendor and, when the hot site is needed, pay an additional "declaration fee"—to declare a disaster and reserve a system and space ahead of other possible claimants. The company brings its latest backup data tapes (has them shipped or electronically transferred to the site), loads its programs, and resumes operations at the hot site. Daily-use fees are usually payable as long as the hot site is occupied.

Because each site is limited in its capacity, employees may be reluctant to travel great distances from their homes to occupy the hot site, especially after a large disaster. Depending on the potential impact of an outage, the use of an internal hot site may be the most practical solution for the rapid recovery of data systems. Duplicate data centers are maintained in different company locations, with all transactions of the main center immediately mirrored (duplicated or replicated) on the alternate system.

Recovery plans should detail the step-by-step instructions required to transfer operations to the hot site and list the employees who are to occupy the site. Most hot-site vendors have personnel on staff to assist with developing the transfer plan.

A *cold site* consists of an empty facility or leased space where computer hardware, telecommunications, and furniture would be delivered to construct a temporary processing capability. At a cold site, nothing is prewired or ready for immediate operation. Obviously, this is a less expensive strategy, but because of the time required for setup, it may not be a practical solution. Something in between a hot and cold site is a *warm site*.

Relocation

Another common strategy is to simply relocate from one part of a damaged building or site to another. Executive suites, hotel rooms, client and vendor offices, empty warehouses, or mobile home trailers are other options to consider to relocate some or all of your business functions. The use of circus-type tents is generally not a good strategy.

Work at Home

Many employees, given the proper resources ahead of time, can work effectively at home. This may free office or work space for those who can't.

Telecommunications

Even without widespread damage, phone systems, including the cellular system, can become overloaded and inoperable. Many of the strategies used to recover data systems are also used for telecommunications. These include emergency service and replacement agreements, divergent routing, radio systems (radio frequency and microwave), mobile switches, third-party call centers, and hot sites. Most switch vendors offer emergency service and replacement agreements. Divergent routing should be examined closely with a carrier representative. Cellular telephones add some degree of redundancy for low volume or emergency calls. Mobile satellite transmission can be used as a backup or as a form of diverse routing. Microwave transmission is a method to add redundancy to connections between buildings of a campus or across town. Lower-frequency radio and infrared transmitters are also used for network connectivity of desktops and servers. This will save the cost and time of recabling networks after a disaster. The simplest way to ensure the continuity of inbound communications is to transfer all calls to another company location if the equipment can accommodate the extra volume and if there are a sufficient number of knowledgeable operators to answer the calls. Commercial call centers are available to handle overflow traffic or to act as a substitute for your operators. Most call centers operate like data center hot sites, with similar fee structures. Their operators can take messages, forward calls, explain the situation, or if qualified, take orders and answer technical questions.

Third-Party Manufacturing

Many firms in the United States and Japan were directly affected by the Kobe earthquake because their only sources of raw materials or parts were from that region. Inability to get parts required these companies to reduce production, find and qualify alternative suppliers, order new parts at inflated prices, and suffer through delayed delivery schedules. The solution: Identify sole-source suppliers and take action to find alternatives far ahead of such problems. If the operation uses "just in time" manufacturing, arrange to warehouse a sufficient quantity of material to allow for delay caused by a disaster or contingent interruption. Some distributors will warehouse materials at your location, retaining ownership until the material is removed and used.

Another concern is the loss of manufacturing equipment, facilities, or personnel due to labor action, inclement weather conditions, or natural disasters. If it is not feasible to transfer operations to other locations within the company, make arrangements with a contract manufacturing firm to produce or assemble your product. As with reciprocal agreements, make these arrangements ahead of time and forward all production change diagrams to the vendor as they occur. Consider using these vendors to supply a small portion of your regular production to check quality, reduce ramp-up time, and familiarize the vendor with your operations and expectations.

Purchase of Materials from Competitors

A manufacturing plant recently destroyed by a tornado had no redundant processes or viable alternate sites. Their expected time frame to rebuild was 6 months. The company was certain to lose long-standing customers if delivery schedules were not met. Their solution was simple. They purchased their competitor's product at a slightly higher price, relabeled it, subjected it to stringent quality tests, and shipped it as their product with a note explaining the circumstances. Not one customer was lost.

Data Systems

Data recovery strategies include hot sites, spare or underutilized servers, the use of noncritical servers, duplicate data centers, replacement agreements, and transferring operations to other locations. Data policies and procedures will help to prevent "disasters" caused by users. To recover data systems, identify the critical applications and prioritize the order in which they are restored. If applications or operating systems are dependent on others, restore them first. Once the applications are prioritized, identify where these applications reside. This will tell you which server or system to recover first.

Servers that are on the same network (or can be easily connected) and that have excess capacity can be pressed into service to rescue a server that has failed. Some organizations keep spare, preconfigured servers in storage for immediate replacement if a primary fails. Unfortunately, this is a very costly strategy.

Duplicate systems—capable of processing normal operations, installed within the organization, and used to run test programs or other noncritical processes—can be pressed into service if a main system fails. Few managers, however, can justify the expense of such duplicate systems. The supply of commercial hot sites is limited and could easily be saturated in a regional disaster, leaving the organization without a recovery system or location.

Revert to Manual Methods

More and more functions rely on automated systems to perform their work. When the automated systems fail, businesses can revert to the manual methods used before the system was automated. For example, a mail-order electronics distributor types an order into a form that resides on a server. The server sends the "pick list" to the warehouse, deducts the item from inventory, and sends a report to accounting after it has billed the customer. If the server fails, the person who takes the order fills out a three-page NCR ("no carbon required") form and physically sends a copy to the warehouse, inventory control, and accounting. When the server is repaired, automated methods resume, and temporary employees are brought in to input the NCR forms into the system.

Unfortunately, with high turnover in many organizations, few employees remember how the job was done before automated methods were used. Often "new" manual methods are developed, and recovery teams are trained in their use. Many high-technology companies, however, cannot use manual methods to manufacture their products, and many processes in use today cannot effectively use manual means.

Virtual Manufacturing

Some companies have the ability to use virtual manufacturing if a production line or facility fails. This strategy worked so well in one case that the company didn't bother to restore its assembly line. Agreements, assembly diagrams, and data connections must be established before the disaster.

Workforce Management

Working extra shifts with the existing workforce or with temporary personnel is a simple strategy to recover from a short-term outage, especially when employees are cross-trained to perform a variety of functions. Decide what functions can be suspended and if employees from those functions can be borrowed. Every continuity plan must consider human resources issues. During recovery operations, ensure there is plenty of food, water, comforts, and rest for the recovery teams. Schedule all employees so that they do not work more than a 12-hour shift. Bring in masseuses for the management staff and other team members. Keep psychological counselors who specialize in post-traumatic stress on call and arrange brief 10- to 15-minute individual meetings with all employees. Those in need of additional counseling can be best identified in this manner.

Reciprocal Agreements

Excess capacity at other sites, similar industries, or even competitors can be used to remain in production until damaged facilities are repaired or replaced. The protection of proprietary information, disruption of the host's operations, and fluctuations in the amount of excess capacity can make this a difficult strategy.

Equipment Rental

If equipment is damaged or destroyed, many plans call for their temporary replacement with rentals. List this equipment and its sources in the plan. Whenever possible, have the rental company preconfigure the equipment to your specifications. Remember that other firms may be after the same equipment, so have alternate or out-of-town sources available. Arrange for priority agreements when possible.

Rescheduling Production

A priority task for many companies after a disaster is to determine the expected length of the outage and compare this to remaining capacity, current production schedules, critical deadlines, and pending product releases. Decide whether production schedules should be changed to concentrate on the most critical products or to eliminate others.

Reallocation of Resources

Similar to rescheduling production, firms should reexamine the assumptions, strategies, and critical time frames and compare them to the extent of the disaster. As necessary, reallocate resources among teams, functions, or sites.

Service-Level or Quick-Ship Agreements

The destruction of a building full of desktop computers would represent not only a monetary loss of equipment, work in progress, and possibly the data residing in the computers but also a major delay in recovery—because of the need to purchase, deliver, set up, reconfigure, and reload each computer. Once computers are installed and connected to the network server, reinstallation of the applications can be accomplished somewhat automatically. But even this, if it is possible, could require a lot of time. If the loss of the equipment is the result of an area-wide disaster, you will be competing with other large companies for replacement equipment. To avoid these delays, you can enter into agreements with computer manufacturers or third-party suppliers to deliver large numbers of preconfigured computers within 24 hours to your primary or alternative location. The same applies to server repair and replacement. Your applications, configured to your environment, are installed by the vendor before shipment, saving you valuable time and resources.

8. Set Up Recovery Teams

Some organizations rely on a single team of executives and key employees to direct recovery operations after a disaster. Sometimes referred to as the *crisis management team,* it decides what individuals or departments within the organization will do to effect the recovery. Their decisions may be based on detailed preplanning or on a loose set of recovery or continuity strategies.

A more effective method involves the formation of individual recovery and continuity teams arranged along departmental lines or drawn from several departments with similar functions (and therefore with similar recovery strategies). Large departments or teams may contain support teams, or subteams, that focus on particular functions or resources.

Each team is composed of a leader, an alternate leader, and essential personnel. The reporting hierarchy extends to a management team through the continuity coordinator (Figure 14.1).

This structure allows for a response that is selective (not all teams need to be activated in every recovery situation), coordinated (information flows efficiently up, down, and across the recovery organization), and focused.

A corporate team may exist in organizations with many divisions or multiple geographical locations. The corporate team is responsible for making strategic business

FIGURE 14.1 Continuity Team Structure.

decisions and will direct the recovery process on a regional basis. Team members will include top management and the business continuity coordinator. Other responsibilities may include the following:

- The safety of all personnel
- Assisting the site disaster recovery management team to decide whether an alternative work site is required
- Projecting and tracking the financial impact of the disaster
- Determining the need to review the strategic position of the company based on any change or expected change in financial position, production capacity, corporate image, or sales
- Working with the public relations director to develop messages and positions, and communicating the necessary management decisions to the public relations team—that is, activating the crisis management plan
- Resolving conflicts with the allocation of resource requirements among multiple sites affected by the disaster
- Ensuring that insurance claims are filed in a timely manner
- Monitoring the recovery operations and recovery expenses
- Keeping the board of directors updated on the position of the company and on the progress of the recovery operations
- Monitoring and assisting the site disaster recovery management teams, local recovery coordinators, or facilities teams with building restoration, relocation, and the acquisition of temporary or permanent replacement facilities

The site business continuity management team is responsible for the coordination of the continuity efforts of the local teams. In smaller organizations, it may assume many of the duties of the corporate team.

One member of the disaster recovery management team is designated as the continuity coordinator. The continuity coordinator is responsible for the overall operation of the recovery. The continuity coordinator activates teams as necessary if they have not self-activated and acts as the liaison among team leaders, the disaster recovery management team, and the corporate team.

The team leaders activate their plans and notify their team members. They are responsible for overseeing the implementation of their teams' recovery instructions. The qualities of a good team leader include the ability to take charge in an emergency situation, familiarity with the operations of the functions to be restored, and freedom from other significant recovery duties that may interrupt focus. For political reasons, the department manager is most often selected as the team leader.

Only employees "critical" to an operation are selected for the recovery team. "Nonessential" personnel are assigned to other duties or to other teams as needed, or are temporarily furloughed. In today's business environment, where staffing is lean, fewer employees (and business functions) are considered nonessential. The duties imposed on the team members by the recovery instructions must closely match the members' normal skills and scopes of responsibility. If members are to perform special functions outside their normal duties, they should receive continual training in these new skills beforehand.

Some recovery planners advocate organizing the business continuity planning structure according to the Incident Command System (ICS; see Chapter 13, Response Planning). Although ICS is an appropriate method to manage a field response to an emergency, I do not believe it is effectively adaptable for business continuity planning. Future regulations, however, may require its use.

Steps 9 to 11

Guidelines to develop team recovery instructions, collect resource information, and document the plan are included in Chapter 16, Plan Documentation. After these tasks are completed, it is time to train team members, validate the plan, and keep it up-to-date.

12. Train Recovery Teams

All employees are trained in some aspect of the plan, even if it is to simply make them aware of its existence. The planning process should accomplish most of the orientation and training required to implement the plan. Those with an active role in the recovery should understand all aspects of their duties and all components of the plan. This includes the methods the company and employees will use to communicate with each other and their responsibilities after a disaster. The importance of record keeping, lines of authority, and team structure must be emphasized. If the duties of team members differ from their normal responsibilities, they should be well versed in these new skills beforehand. An orientation is a good way to introduce the plan to the general employee population. This can be combined with home disaster preparedness training and disaster fairs. Workshops, videotape presentations, interactive CD-ROM, and intranet pages are effective for training targeted to specific levels of plan responsibility.

13. Exercise the Plan

No plan is complete until every element has been subjected to some type of testing, exercise, or simulation. Simulating the plan will validate the effectiveness of strategies, ensure the accuracy of information, and increase the preparedness of the individuals who will execute the plan. It will pinpoint areas that need attention or improvement and reveal gaps in instructions, misplaced or absent assumptions, or the need for better strategies and tasks.

The term *testing* is no longer used because it connotes a pass/fail mentality; most planners believe their overall efforts are best served by promoting a positive outcome, and therefore they offer better motivation for the participants. *Simulation* is probably a better term because to many of us, exercise is not something done willingly. Still, some old-timers like the term testing because it places some degree of stress on the participants, thus creating a more realistic situation.

Many publications exist that guide the reader through the steps of a tabletop, departmental, functional, and full-scale simulation. A complete treatment of exercise planning is beyond the scope of this book because the planning for a full-scale exercise involves (or should involve) a significant effort. All portions of the plan must be simulated and contact information verified. A simulation must be planned in detail if it is to be effective and not disrupt normal operations. Begin simply by simulating individual business unit plans. Simulating the entire plan at once is a major event that can take 6 months to set up. Starting out with a small piece maximizes the chance for its success. Later, as your exercise program matures, include city, county, or state agencies in a full-scale simulation. At a minimum, plans should be simulated annually. Three levels of exercises are tabletop simulations, functional exercises, and full-scale exercises.

A basic or first-level exercise is a "walk-through" or tabletop simulation. A tabletop simulation is discussion with the continuity team members executing their plan based on a scenario. The team is given the scenario and asked to apply or walk through their instructions. The continuity coordinator or manager can act as the controller who both runs the exercise and keeps it moving. The controller can change the parameters of the scenario as needed, but should not do so unless he or she sees the need or advantage to go in a different direction to stimulate discussion or revise assumptions or resources. The controller, at the beginning of the exercise, makes it clear to the team members their purpose, the rules of the exercise (usually none), and the fact they are not there to find fault, but rather to validate their strategies and tasks. Someone should be assigned responsibility to record the minutes of the exercise, noting what went correctly, what did not, and what action items result. Ideally, the scribe should not be part of the team. As the controller, be prepared if this is the first time the team has actually read the plan.

Scenarios must be believable, realistic, and relevant. An asteroid striking the Houston Astrodome will put many to sleep. The scenario must present a situation that causes an outage that tests tasks and resources and that satisfies the objectives of the test. Present the scenario and let the team leader execute the plan.

A higher-level or next-step tabletop simulation can involve an exercise with one or more dependent teams to test their ability to coordinate tasks, provide input or output to each other, or uphold communications. Tabletop exercises are also used to test security plans and procedures and are more appropriately (usually) threat or event based.

Another simple but useful exercise is a "call-tree" simulation, whereby team members are contacted by the various means outlined in the plan. Some companies issue pager, Blackberry, or other communication devices. Each team member is contacted and told to report a special code number at a convenient location the next morning to receive coffee and pastries or the like. Those who did not receive notification and those who don't report are queried about the reason.

Yet another exercise involves traveling to the alternate work site, especially if it is within driving distance, to familiarize team members with its location, main and alternate routes, and check-in procedures. Surprisingly, some team members may have difficulty finding the location even if just 30 miles away!

The relocation exercise can be combined with a functional exercise that actually tests the ability to physically accomplish a task. Many team members are issued laptop computers to log onto a different data center from the alternate work site. After the team relocates, members are required to log onto the site. A functional test can also include switching data processes to an alternate data site, testing communications, or similar activity.

A full-scale test can include actually switching data systems or restoring data at a hot site and alternate work site. Other types of full-scale exercises can include drills with city, state, or county emergency services or the activation of the Emergency Operations Center and the management team. This type of exercise is more involved and can take 6 months to plan. The scripts are a little more complicated, with multiple controllers and observers involved. Whenever the management team is asked to participate, it must be well conceived and executed.

To conduct an exercise:

1. Decide what to simulate.
2. Select objectives.
3. Develop a realistic scenario to test objectives.
4. Select test controllers.
5. Select test observers.
6. Write scripts for simulation volunteers and messages to introduce new situations into the simulation (full-scale exercises).
7. Distribute memo to participants describing what, when, and where.
8. Conduct the simulation.
9. Discuss the results and prepare a report.
10. Revise the plan based on the lessons learned during the exercise.

Limit the number of major objectives to no more than six. This will help keep the goals focused and uncomplicated. A secondary goal of a simulation is to enhance training. This is more effective when the objectives are limited.

14. Maintain the Plan

These plans must be "living documents." Employee and vendor contact numbers change often and must be kept current in the plan. This information must be reviewed quarterly. The plans must be reviewed annually to determine whether they still match the overall strategic direction of the organization, and be changed accordingly.

In large organizations, the team leaders are generally required to keep their plans up-to-date, with their efforts audited by the recovery coordinator or by internal audit.

The results of the audit, that is, the freshness of their plan, are reported in score card format to upper management. Many organizations try to interface their employee database with that of Human Resources, so that any changes made in one updates the other. Termination causes a flag that a team member needs to be replaced. Others are allowed to update their own information on line.

REVIEW

Business continuity planning is a process that identifies a company's critical functions, develops cost-effective strategies to recover those functions if they are lost or if access to them is denied, and lists the instructions and resources necessary to implement the strategies. Systems, applications, products, and processes are prioritized and recovered in a logical manner that will allow the firm to remain in business and to retain or gain market share over competitors that don't have a continuity capability.

It is in the planning process itself that the true value of continuity planning is realized. Although some planning methodologies are more effective than others, each must be adapted to fit the corporate environment and culture. In most companies, the planner will not be successful without the committed support of top management.

Business continuity consultants can be used to facilitate the project and advise the company on the best recovery strategies. The more common strategies include data backup; prioritization of systems, applications, functions, and equipment; and transfer of these operations to an alternate location. Strategies and plans must then be devised to repair, rebuild, or relocate the business. Do not forget to devise plans to move out of the alternate location and back into a permanent facility.

15

Business Impact Analysis

Eugene Tucker, CPP, CFE, CBCP
Contributing author

"The great enemy of the truth is very often not the lie—deliberate, contrived and dishonest, but the myth, persistent, persuasive, and unrealistic. Belief in myths allows the comfort of opinion without the discomfort of thought."
—John F. Kennedy

A business continuity plan that is not predicated on or guided by the results of a business impact analysis (BIA) is at best guesswork, is incomplete, and may not function as it should during an actual recovery.

The BIA will help the company establish the value of each business unit and business process as it relates to the organization and not to itself, illustrating which functions need to be recovered and in what order they are recovered. It identifies the financial and subjective consequences to the organization of the loss of its functions over time, highlights interdependencies, and establishes the function's "outage tolerance" or recovery time objectives (RTOs). Its results are used to determine which functions are the most critical, at what times they are critical, and which strategies are the most cost-effective. It is fundamental in the understanding of the amount of risk to retain, transfer, or mitigate and will assist management to make timely decisions about future business issues. This is accomplished by examining impacts over significant blocks of time (hours or days) on service objectives, cash flow and financial position, regulatory requirements, contractual issues, and competitive advantage. It presents management with a financial basis for selecting the most cost-effective recovery strategies. A risk analysis and a BIA can greatly reduce the cost of insurance by identifying and quantifying a potential loss, thereby allowing the risk manager to avoid overinsuring or underinsuring the risk.

A BIA will also help to accomplish the following:

- Identify which processes and computer applications are critical to the survival of the organization
- Identify critical resources of the organization
- Gain support for the recovery process from senior management
- Increase management's awareness of the issues and resources required for a workable program, as well as introduce a basic planning structure to the management group
- Potentially satisfy regulatory requirements and standards

- Potentially reveal inefficiencies in normal operations
- Help to justify or allocate better recovery planning budgets (cost/benefit)

A major objective of the BIA is to establish the RTOs and recovery point objectives (RPOs) of business functions and data processes. Outage tolerances or RTOs are established based on the results of the analysis and the recovery priorities that are assigned. Management will then understand how best to allocate recovery resources to these functions. A shorter RTO will require the most expensive strategies; a longer RTO will allow for the selection of lower availability options. Outage tolerance is the amount of time the organization can be without the use of a function before it has a detrimental effect on the company. The recovery time objective is the amount of time by which the organization would like to have the process or function back in service. These terms are often synonymous. The RTO is the maximum downtime before the function or process is critically affected or the deadline in which the function or process must be restored to prevent severe impact to the business.

The RTO is a decision based on the results of the BIA and on the judgment and agreement of the process or business unit owner with the concurrence of senior management. The definitions refer to critical or severe impacts, but the organization can make its own decision about what level of loss or impact is tolerable. Because major recovery strategies are dependent on and should be driven by the RTO, senior management acceptance is vital to the funding of the strategies.

RTOs can also assist in the escalation of a problem to the declaration of a disaster or to a higher level of response. If the time required to recover a process is 4 hours and the RTO is 6 hours, you have about 2 hours to fix the problem or to decide what to do about it.

The RPO is the point in time to which systems and data must be recovered after an outage (disaster recovery journal) or the acceptable amount of data that can be lost before a function or process is critically affected. In other words, how much data loss are you willing to accept in your recovery, or how current does your data need to be to start your recovery? It may be acceptable to lose 4 hours of data, and then recover to that point, recreating the data that was lost. The RPO is often used in the selection of a backup strategy. Tape backup could mean a day or more of lost data.

When developing the business continuity plan, keep in mind that a data process that supports the RTO of a business function may need to be recovered sooner. If the function cannot begin recovery until the data are ready, that time differential must be considered in the dependent RTO and RPO.

Another objective of the BIA is to show the business cycle criticality of various functions and refine strategies or reallocate resources accordingly during a continuity situation. Companies are composed of individual business functions that work together to deliver services or products. Although all functions are important during normal operations, some are more critical or time dependent than others during a recovery situation. In most cases, not all business processes need to be recovered at the same time. Functions initially less critical can become more important over time, and some functions can have more importance depending on the timing of the business cycle or retail season. Companies that survive after a disaster do so by focusing on the recovery of their most critical or time-dependent functions, by implementing the most cost-effective recovery strategies, and by making the best use of resources that become strained or scarce after a disaster. Be careful that the prioritization of functions is not based solely on their outage tolerances.

Their relationship to the organization's mission is the most important factor. Impact information is also used in the cost/benefit analysis of recovery strategies.

It is obviously important to decide which recovery strategies to use before scripting recovery instructions or steps. Often, the selection of a recovery strategy necessitates a change in policy or even the strategic plan of the company. Management may have difficulty making these decisions if the impact, solutions, and costs are not presented in a complete, positive (avoid a "savior versus doom-and-gloom") package.

A purist will advise the planner to treat the BIA as a project separate from the continuity planning phase. He or she will argue that the planner's only objective at this point in the process is to demonstrate the cost (impact) of a loss over time. This makes sense, in that functions are identified as critical and are prioritized in large part by their financial impact. As a practical matter, however, this is difficult, and it comes with some disadvantages. The BIA will generate interest, support, and momentum in the continuity project as a whole. These advantages will diminish over time, so the planner must complete the initial phases of the project as quickly as possible. Separating the BIA from other elements of the continuity planning process will often add unnecessary time and expense, which can result in failure to complete the project. Management time is often at a premium; thus, it is difficult to arrange more than one meeting to conduct the impact analysis, identify critical functions, and discuss recovery strategies and resource requirements. This is less of a problem in a large organization if the planner takes a top-down approach and has the opportunity to examine issues of mitigation, preparedness, response, and recovery at a lower level of middle management. Unfortunately, within most organizations, the best source of this information often resides only at this lower level. The planner may have only one opportunity to meet with these individuals. As a result, planners, especially those who use questionnaires extensively, include questions related to mitigation, preparedness, hazard identification, and resource requirements as part of the BIA meeting. Critical functions, the identification of single points of failure, and the initial selection of recovery strategies are often discussed at this juncture.

RISK ANALYSIS VERSUS BUSINESS IMPACT ANALYSIS

A BIA is often thought of as another name for a risk analysis. Some contingency planners believe a risk analysis is a process that focuses solely on physical assets and that a BIA focuses solely on business processes. A close examination will reveal that this belief is not correct.

A BIA is a means of assessing the impact of a disruption in any functional area or on the operations of the enterprise as a whole. It can be considered a subset of a risk analysis, in that it places an "asset value" on business functions and focuses on the criticality of a disruption over various time periods. The source or cause of the disruption, or a detailed understanding of the probability of its occurrence, is relatively unimportant when conducting an impact analysis and is not considered.

Therefore, the BIA focuses much less on hazard identification (some say there should be no focus on hazard identification). Listing all the hazards that might befall an operation may be useful in understanding the conditions, environment, and special needs required when selecting recovery strategies (such as the inability to move large pieces of equipment across town after an earthquake or the amount of time required to remove bodies, complete an on-site police investigation, and remove the carnage after a workplace homicide), but it adds little to the understanding of financial or subjective loss to the operation over time. Outage tolerance is exclusive of its cause.

The approach taken to initiate and manage a BIA is very similar to that used in a risk analysis. It must begin with senior management's commitment. Because a top-down approach provides the fastest and often most accurate results, management must emphasize that this is a task not to be delegated downward in the organization. The planners' discussions with these managers in the process of collecting impact data will help to build relationships that will be useful later on if the business continuity planning process becomes stalled, or if resistance is encountered.

The risk analysis is a more or less solitary experience, whereas the BIA is a top-down partnership with senior management. Advantages of this partnership include the following:

Additional Interaction with Upper Management

If the planner or security manager is at a lower level, bound by a tight reporting structure (as is often the case), partnership with senior management is an exceptional opportunity to familiarize oneself with their concerns. It allows senior management to recognize the value of your department and you as its leader.

Acceptance and Validity

The data collected in this way are more often accepted. Data presented to management by an outside consultant or developed by a bottom-up approach may be met with skepticism and subject to increased scrutiny. Unless the consultant or you are intimately familiar with the financial position of the company, either may have difficulty responding to senior management's questions in a meaningful way. When the analysis is developed using a top-down approach, the management group will have had a hand in developing the data and will tend to answer each other's difficult questions themselves.

Support

The partnership with senior management should generate the support needed to drive the project through the remainder of the process. Without this support, continuity planning tends to bog down, die, or take so long that a great deal of time must be wasted revising information that has become outdated before the plan can gain acceptance.

Meaningfulness

The risk-matrix approach may be too simple and have little meaning to senior management if the results of the analysis are not stated in financial terms. Statistical probabilities can be confusing, and they tend to present a doom-and-gloom feeling. If statistics are used, they should be simple and accurate. Although some codes and standards suggest the use of a matrix approach, bear in mind that some outline minimum standards. We believe that BIA or other methods that put financial data into the equation should be above this standard.

BUSINESS IMPACT ANALYSIS METHODOLOGY

Plans based on intuitive analysis are often too generalized or miss details important to an effective recovery. A comprehensive analysis often reveals interdependencies and outage tolerances that are not obvious even to those with intimate knowledge of the company's operations. The BIA, if conducted in a structured manner, can help guarantee the success of the entire business continuity process. Different methods exist to accomplish this task. Whatever method is used, certain steps must be completed if the planner is to avoid obstacles commonly encountered in the analysis. The major elements of a BIA include the following:

- Project planning
- Data collection
- Data analysis
- Presentation of data
- Reanalysis

Project Planning

The following initial steps should be taken in the development of a BIA:

- Management commitment
- Definition of the scope of the analysis
- Identification of the participants
- Deciding how to collect the data
- Arranging interviews or distributing questionnaires

As stated previously, the biggest single predictor of the success or failure of the business continuity planning process is the level of senior management commitment. Most often, this commitment is gained only after management becomes fully aware of the potential harm to the company from the loss of its ability to deliver service or products. These losses can be fully demonstrated through BIA. The BIA owes a great deal of its success to senior management support!

But which comes first, and how? Planners who have the ear of the chief executive officer (CEO) or an influential board member have little difficulty getting the needed support. Unfortunately, most planners are not blessed with this degree of leverage. To get the process started, the planner must find or convince someone in senior management to sponsor the project. Often, this is the chief financial officer (CFO). Comprehensive financial audits increasingly mention the lack of a recovery plan. The CFO is more readily convinced of the importance of the project, owing to his or her intuitive understanding of potential financial impacts on the organization. This is the person who also may be legally responsible for the protection of certain corporate documents. The CFO is usually high enough within the organization to drive the project and to help the planner both collect and interpret the data. If the planner has difficulty getting financial data from other managers, a "down and dirty" analysis can be derived from the CFO's knowledge alone.

Because the BIA examines a great deal of financial data, the planner would normally maintain a close relationship with the CFO. The results of the analysis must be acceptable to management—everyone must believe that the analysis represents the true impact to the organization. The CFO should provide good insight into how the data in the final report should be represented.

The CFO can convince the CEO and other senior managers of the need to commit themselves to the project and of the value of a top-down approach—don't give senior managers an opportunity to delegate this task downward within the organization. Beginning the analysis at this level will help to ensure the accuracy of data by avoiding the tendency of department managers to inflate the importance of their business units or functions. Lower managers usually do not have the "big picture" of the impact that loss of their functions could have on the objectives of the organization.

The BIA should be completed in the shortest time possible, in order to produce the most accurate results. The financial position (impact), even the organization of the company, can change rapidly, and data may become outdated if more than 3 months old. Consider the use of outside consultants to assist with the interviews and analysis if the scope of the project is sufficiently large to justify their use. The interest and momentum gained for the recovery planning process will be diminished if this phase is extended for any length of time. Enlist senior management's help in arranging to have all participants available for the study. A senior executive or the project sponsor should issue a memo to all affected personnel describing the importance of the project and introducing the project leader.

Senior management should agree on the scope of the analysis, timelines for the project, type of outcome expected (for instance, the format of the presentation), and who is to participate. Product introductions, or a pending Food and Drug Administration inspection in a biotechnology firm, are not times when full cooperation can be assumed. You will have better success when all participants are otherwise available.

Normally, the scope of the analysis will mirror that of the business continuity planning project, but management or the planner can tighten the scope to an individual division, site, or location. Many planners make the mistake of including in the analysis only those departments or business functions they believe are critical without the understanding that a major goal of the BIA is to determine what is, and what is not critical. At this juncture, assume that all functions are critical at some point in time, or under certain circumstances, and include them in the analysis.

Meet with the CFO or your sponsor to determine who should participate in the analysis. Interview the highest-level manager in each functional business unit. Try to maintain consistency among the positions of the participants; that is, interview only at the director or vice president level. Organization charts are useful tools to guide the planner to the proper levels within the organization for interviews. As with continuity planning, many planners insist that a planning group, composed of the planner and members of management, must develop questionnaires and determine what impact the organization will suffer from disasters. They believe this is the best and only method to ensure management support and an effective planning process. In fact, planning groups can be useful, if the time allotted to do the analysis is small or the number of business units participating is very large.

Prepare an informal list of potential financial impacts that may affect the operations of the company (see Box 15.1, Sample Questions). In the meeting with the project sponsor, review the types of questions and assumptions you intend to use. Discuss any additional impacts or risks the sponsor may wish to add to the list. If you are a consultant, the sponsor can ensure that terminology used in the questionnaires is consistent with that used in the company because this will help to avoid confusion or inconsistencies in the data.

Determine how best to manage the collection of the data. Data collection is accomplished through the use of interviews, questionnaires, or both. In large organizations, a workshop is sometimes conducted to outline instructions and explain expectations. Questionnaires distributed directly to managers tend to be delegated, so avoid their exclusive use unless they are to be completed during the interview. Questionnaires are best used in a bottom-up approach to collect resource information that you will incorporate in the plan.

Schedule interviews. Give a brief presentation to the participants outlining the purpose of the analysis and giving them an opportunity to collect the information you need. It can be difficult to get managers to think in terms of the financial and subjective impact of their operations and to discuss recovery strategies, especially if they believe they will be responsible for selecting the best strategy then and there. This belief may overwhelm some managers and cause them to mentally give up on the project before it starts. Avoid this problem by clearly explaining to the manager the goals and expectations for the project before the interview.

In addition to the CFO, the risk manager is an important resource for the analysis. Meet with the risk manager to review insurance requirements. Determine what perils and property are excluded in existing policies. Ascertain the period of indemnity for business interruption and contingent business interruption, and estimate the time delay in reimbursement. Calculate the value of the physical assets and identify lead times for the replacement of assets. These values are often not considered because they are subject to insurance reimbursement; however, it is important to know whether the policies call for reimbursement at replacement or depreciated values. Although we will not reduce the financial impact in the analysis by the amount of reimbursement, we can add the difference in the replacement versus depreciated cost. The loss of interest on cash reserves or on loans to secure temporary or replacement property, and extraordinary expenses during the delay between purchase and reimbursement, should also be added to the calculations later in the analysis.

Although it is more effective to concentrate on the restoration of critical business functions than on individual computer systems and applications, as was common in the past, critical functions can include "mission critical" processes, equipment, and applications. It is extremely important at this point to also meet with the information systems director to determine the impact of losing these functions as well as other vital communication links within the company.

Data Collection

To obtain data that are valid, examine all current business functions and operations. Some planners suggest that only critical business functions be analyzed, but again, the purpose of this analysis is to determine which functions are critical. Any designation of a function, operation, or process as critical or noncritical before the analysis is merely intuitive. Intuitive estimates of potential acceptable outage tolerances or downtime can misdirect many thousands of dollars to recovery strategies that cause overplanning or unnecessary loss during a recovery situation. Data for the analysis are best collected through interviews with unit or department leaders. This approach helps to gain their buy-in for the project more effectively than if they were simply to respond in a questionnaire. During a personal interview, managers can ask questions and better understand what is expected from them.

Start the main part of the interview with the difficult financial and subjective impact questions, so that neither the planner nor manager feels the need to rush through the process as the meeting time begins to expire. If planners cannot schedule sufficient time to cover these topics adequately, they should consider postponing meetings for another time. Allow 1 to 1 ½ hours for each meeting, longer if nonimpact issues are discussed or if the manager is not experienced with putting impacts in financial terms. Although this is a top-down data collection procedure, it may be necessary to meet later with the next-lower level of management to validate impacts and strategies developed.

During the interview, determine the following:

- How the business unit fits in with the overall mission statement of the organization
- The primary service objectives of the business unit
- The business unit's processes and dependencies

The analyst will also guide the manager to accomplish the following:

- Estimate the maximum loss if the function is out of service, out of operation, or access to it is denied for 30 days or more. (This 30-day period can be adjusted to fit expected outages, but it should remain consistent for all business units being evaluated. Assume that the function, or access to it, is lost at the worst possible time of the month, year, business cycle, or production schedule. Be careful that evaluation of functions is not duplicated.)
- Determine how this maximum loss is allocated over the following times: day 1, day 2, day 3, day 4, day 5, week 2, week 3, and week 4. Again, these times can be adjusted to fit the specific environment. For example, it may be more realistic for a financial institution to track its losses by the hour instead of by the day. Consult with the CFO to decide what time periods are best for the calculations.
- Estimate any extraordinary expenses the unit may incur if the function is lost. As before, indicate the maximum amount, as well as the details over a specified time. Extraordinary expenses can include costs associated with idle staff, wages paid to extra staff to handle backlogs, equipment rental, outside services, and transportation.
- List the impact on any contractual or regulatory obligations the outage will cause. What fees, fines, penalties, or missed milestone payments will the company face?

- Indicate how cash flow will be affected by the outage.
- Rate the loss of goodwill or damage to the corporate image.

When seeking answers to these and other questions, be careful not to put participants in a defensive situation. Form questions in such a way as not to cause managers to believe they need to justify their positions within the company. Don't ask, "How valuable is your function to the organization?" Instead, ask:

- What would be the impact to the organization if this function were lost?
- How would this impact change over time?
- How would the loss of this function affect other functions within the organizations (that is, what other functions are dependent on the input to or output from this function)?

Some planners and software developers believe the best method for completing the analysis is a simple distribution of questionnaires to the appropriate participants. Although many planners use questionnaires exclusively, questionnaires without personal interviews will not maximize the planner's understanding of how each function fits into the overall organization and how its interdependencies relate to the whole. Managers are reluctant to put sensitive financial and other impact information "in writing" and may need additional time to answer the questions adequately. Face-to-face interviews can stimulate managers' understanding of the process and, more importantly, their ability to think through strategies and resource needs. A simple compilation of financial information from questionnaires will miss subjective information that could be important to an accurate analysis of the data.

During an interview, the analyst can ask the questions listed on the questionnaire, with the exception of items that are simply resource oriented. Have a printed or electronic (diskette, shared drive file, or web) version of the questionnaire available to give to the manager after the interview. Many of the business continuity planning and BIA software programs on the market have this capability. The manager fills out the questionnaire electronically, and the analyst simply imports the data into the software, avoiding the need to input data twice.

If a questionnaire is used without an interview, meet briefly with the recipients and explain both the purpose and importance of the project. Recipients rarely complete all sections of questionnaires, necessitating follow-up questions. Questionnaires are impersonal.

Box 15.1. Sample Questions

The following are sample questions the planner should ask to obtain the type of information useful to a complete and meaningful impact analysis.

1. If your department or function generates revenue for the organization, what are the sources of this income? Sources of income can include:
   - Product sales (list by product lines)
   - Services rendered to outside clients

Continued on next page.

Box 15.1. (*Cont.*)

- Discounts or commissions
- Interest from investments or floats
- Incentives for on-time or ahead-of-schedule completion dates or milestones
- Tax base, if a government agency
- License and use fees
- Maintenance fees
- Other _____

2. In addition to the loss of revenue, what other types of financial impact are realized if your department or function is lost? These may include:
   - Canceled orders due to late delivery
   - Penalties for late payments
   - Regulatory requirements, late filings, and fines
   - Contractual obligations delayed or not met
   - Interest on borrowed funds
   - Wages paid to idle staff
   - Other _____

3. What is your estimate of the total exposure for each item or product if the inability to function or deliver the service lasted for a 30-day period, assuming the loss occurred during the period of its greatest negative impact? For example, if the greatest percentage of sales for your Christmas products occurs during December, use the sales figures from this month to estimate the exposure. If your greatest percentage of sales for swimwear occurs during June, use the sales figures for June to estimate loss of sales potential for swimwear. Combine these two (or more) totals to arrive at the exposure for a 30-day period.

   List the cumulative minimum and maximum loss for the days 1, 2, 3, 4, and 5 of the outage for each exposure. If your organization includes retail sales, list the daily sales volume for a 7-day period. Continue by calculating and listing the exposure for weeks 2, 3, and 4 of the outage. Again, use a time period that is significant to your business. Recovery times are often thought of in weeks and months, but impacts in certain institutions, such as a bank, can increase dramatically in a few hours and days. In banking, the organization may cease to exist long before the end of a 30-day period. These organizations must therefore calculate their losses by hours, not days.

   It is useful to list these figures in table format (Table 15.1), for ease of input onto a spreadsheet or into a database program that can combine and report the totals.

   Exposures may include:
   - Lost sales (total or by product)
   - Delayed milestone payments
   - Other _____

Table 15.1 Example of Business Impact Analysis Input Table

	Day 1		Day 2		Day 3		Day 4	
Exposure	Min	Max	Min	Max	Min	Max	Min	Max
Lost Sales	0	0	5000	5000	7500	10000	10000	20000
Canceled order	0	0	1000	1000	2500	3000	75000	100000
	Day 5		Week 2		Week 3		Week 4	
Exposure	Min	Max	Min	Max	Min	Max	Min	Max
Lost Sales	20000	30000	75000	125000	350000	500000	750000	950000
Canceled order	100000	250000	450000	750000	750000	750000	750000	750000

4. What types of extraordinary expenses are necessary to implement expected or projected recovery strategies for your business functions? These expenses may include:
   - Transportation costs
   - Rent for alternative space
   - Contract services
   - Emergency requisitions
   - Temporary relocation of employees
   - Temporary employees to catch up backlogged work
   - Equipment rental
   - Additional supplies
   - Other _____

   Determine the minimum and maximum extraordinary expenses over time and list them as described in question 3.

5. At what times of the week, month, year, or business cycle is processing or production especially critical? In other words, when can an outage hurt you the most? This information is useful to the management team in making strategic "corrections" or modifications to the recovery plan made necessary by the timing of the disaster. It can illustrate changing priorities and allow decision makers to redirect resources to where they are needed most, and when.

   Critical times are typically the end of each quarter or year, or some significant product-related event. A toy manufacturer may list the months of September, October, and November as its most critical time, preceding the Christmas season. When using a questionnaire, allow respondents to provide free-form answers. Some questionnaires and BIA software programs only allow for the selection of individual months. Although this monthly period may be sufficient for the analysis, it may not be sufficiently focused for recovery purposes. As pointed out earlier, a financial institution may not be able to recover from a month's outage. Later, this information is matrixed in the continuity plan.

Continued on next page.

Box 15.1. (*Cont.*)

6. What other business impacts might result from the loss of your business functions? The financial loss due to reduced customer service or technical support, tarnished public image, or the loss of future business is difficult, if not impossible, to predict or measure. Some impacts may be purely subjective, or difficult to state in direct financial terms. These impacts can have the greatest effect on the survivability of the company, and they should always be included in the analysis. This questions asks respondents to list what these impacts may be. Examples include:
   - Loss of competitive advantage or market position
   - Loss of shareholder confidence
   - Increased liability
   - Decreased employee morale
   - Cash-flow difficulties
   - Reduced public image or confidence
   - Reduced customer service
   - Contractual consequences
   - Regulatory violations and consequences
   - Loss of key personnel

7. How would you rate the severity of these impacts if your business functions were lost? On a five-tiered scale, rate the estimated severity of the impacts if the outage were to occur during the worst possible moment. A value of 1 represents a minor impact, whereas a value of 5 represents a severe or fatal impact on the function or the organization. Any wider range of choices, such as a scale of 1 to 10, adds ambiguity to the evaluation.

   Some analysts attempt to assign specific definitions to each value. Although the definitions vary, the following is typical:

   $1 =$ Minor or no impact (the problem is easily handled, functions are not affected, or it is not an issue)

   $2 =$ Somewhat critical (although the problem is still easily handled, some degradation in the image, service provided by the function, or ability to meet requirements will occur)

   $3 =$ Moderate (nearly all of the functionality or ability to meet requirements is degraded)

   $4 =$ Serious (continuation of the function or ability to meet requirements is extremely difficult)

   $5 =$ Severe (the function or ability to meet requirements ceases)

   The criticality categories listed elsewhere in this publication can also be used to express the severity of the impact. Use these impacts to help to determine which functions are critical to the organization and to demonstrate to management the consequences of the loss of certain functions. The severity ratings are combined by type of impact for each function, and then by type of impact for the entire company. This can be reported in a table or matrix format, but a graphical representation is easier to understand.

8. How much time is required to reconstruct records or backlogged work once your function is back in operation? This question seems redundant to the calculation of certain extraordinary recovery expenses, but it will help determine RTOs. Based on a reasonable or expected duration of an outage, estimate the amount of time required to catch up, considering that normal (recovered) operations may be concurrent.

   Allow the respondent to give an open-ended response to this question because the range of possibilities can be great. For some functions, the time required can be a few hours; for others, it can be a number of years (and yes, you should discuss proper backup procedures in this instance).

9. What short-term and long-term resources will your function require to operate at a minimally acceptable level while returning to normal operations? This question is intended to give the analyst an overview of critical systems and equipment to help determine outage tolerances and interdependencies. The respondents should detail their resource needs on resource forms or questionnaires after the interview. Questions such as this, and others that ask about the criticality of systems, equipment requirements, application tolerances, and dependencies, are also useful to the issues of critical function recovery and recovery prioritization. Keep in mind that software applications and systems (servers) are not critical functions for the purpose of a BIA—they support critical functions. If a server or application is lost, its impact is the loss of the critical functions it supports. Be careful that such factors are not counted more than once in the calculation of losses.

10. What are the lead times for the replacement and installation of the equipment and systems listed previously? This is an important question, one that can drastically affect the selection of recovery strategies. High-technology equipment can have lead times of 6 months or more. Even if the equipment can be quickly replaced, the re-creation of its environment and support systems can take a long time.

OTHER QUESTIONS FOR THE IMPACT ANALYSIS

The following questions are commonly listed on questionnaires or asked during the business impact interview. They can help the analyst maximize the use of time while gaining information for an effective analysis and continuity plan.

- What is the name of your department, unit, or function? Please describe its function. Include an overview of what your department or team must do to recover from a disaster (you will be asked to list detailed instructions later).

- How long can your department, unit, or function be out of service without adversely affecting the overall operation of the company?
- What are the most critical functions of your department or unit? Which of these functions would need to be recovered first, either at an alternative site or reconstructed at the original site?
- What tasks would be necessary to recover these functions?
- If these tasks for the recovery of the various functions are dissimilar, how many separate recovery teams would be necessary to implement these tasks?
- Who would you select as team leaders and alternate leaders for these teams?
- What customers (internal or external) or other business activities would be affected by the inability to deliver or perform your function? Describe the extent of the impact and estimate the amount of time before the impact would affect the customer.
- What other areas or functions of the company are you dependent on to operate effectively, and how would their inability to deliver services or products affect your operations? How would an area-wide disaster affect the physical delivery of these services or products?
- Where can you obtain other required services, products, or raw materials if access to your present supplier ceased?
- Is your function directly involved with billing, collection, or the processing of revenue? If yes, please describe.
- Have any arrangements or agreements been made with your vendors for emergency delivery of critical resources?
- Where would or could the organization go to perform this function if employees were unable to gain access to the current location?
- Can you depend on the resources of similar functions at other company locations? Please state the names of the units, sites, and office locations.
- Indicate the estimated square footage required to house disaster recovery operations for both a short-term and long-term (or permanent) relocation of your business unit. Attach a copy of any special floor plans needed.
- What computer support systems (that is, HP, IBM, stand-alone PC, or LAN) are necessary for the continued operation of your unit? What is the maximum amount of time these support systems could be unavailable before their loss would have a negative impact on the company? Rank these systems in order of priority and indicate what configurations are necessary. Are these systems maintained by the Information Systems Department or by your department?
- Do your critical vendors and suppliers have tested recovery plans in place, and is their recovery time consistent with your needs?

RESOURCE QUESTIONNAIRES AND FORMS

Ideally, the resources required to implement the recovery strategies are best determined subsequent to the selection of the strategies themselves. Most often, however, questionnaires are distributed after the interview to give the respondent time to list the short-term and long-term resources they would need to recover their operations based on only the preliminary discussion of strategies. Resources are selected from the following categories:

- Employees and consultants
- Internal and external contacts
- Customers
- Software and applications
- Equipment
- Forms and supplies
- Vital records
- Other _____

Employees and Consultants

List the names, titles, addresses, and contact phone numbers for the employees assigned to your team. Set up call trees if desired. A *call tree* is when one team member is responsible for contacting a fixed number of team members who in turn have a list of team members they are responsible for calling. Call trees can expedite the notification of large groups of people. List as many numbers as practical (home, office, cellular, pager, fax, and e-mail addresses) and use a consistent format, such as (415) 555-1212.

Not all employees on your staff are generally needed during the initial recovery phase. After a disaster, space and resources may be limited. Identify at what point during the recovery process the individual is required, according to the following:

1 = Required within the first 24 hours
2 = Required after 24 to 72 hours
3 = Required after 3 to 5 days
4 = Remain on standby until advised

If plans are maintained electronically, as most now are, the address section should be separated into street, city, state, and zip code columns so that a database sort can be achieved (Table 15.2). Such action may be useful to determine who lives closest to another or for identifying who lives inside an affected area.

Internal and External Contacts

List the names, addresses, and phone numbers of vendors, suppliers, consultants, or other people and groups, either internal or external to the organization, you may need to contact

Table 15.2   Sample Employee Resource Questionnaire Form

Function:_____

Employee	Title*	Address	Contact Information			Priority	Called by:
			Home	Cellular	Pager		

*Title, position, or recovery function.

Table 15.3   Sample Internal/External Contact Resource Questionnaire Form

Function:

Contact or Vendor Name	Representative	Address	Primary Phone	Alternate Phone	Type of Service	Account Number or Password

to assist with the recovery or with replacement of supplies, equipment, or raw material, or to provide technical support (Table 15.3). This list should be complete but as short as possible. Bear in mind that your original list of contacts (phone books, vendor lists, etc.) may be buried under tons of debris after a disaster.

Briefly describe (one or two words, if possible) the type of service each contact provides and list any account numbers, passwords, or other type of authorizations required. If there are contact people within a vendor company, you may list those names also. If the vendor has a corporate office or other location outside the potential disaster impact area, consider including those numbers also because the local office may suffer damage from the same disaster. Examples of these contacts include the following:

- Plumbers
- Electricians
- Seismic or structural engineers
- Janitorial service
- Security service
- Food service vendor
- Payroll processing
- Hot-site providers
- Portable toilet/shower providers
- Insurance broker
- Bank manager

- Federal or regulatory agencies
- Media contacts
- Equipment/systems suppliers

Customers

Although customers may be listed under "Internal/External Contacts," you may want to list separately certain customers or contacts you would wish to inform of a disaster, in order to let them know of major delays, when you expect to return to normal operations, or that you have relocated to an alternate location. If your organization is unaffected by a regional disaster that has received some publicity, you may need to affirm to these customers your ability to continue to serve them. The questionnaire is similar to the Internal/External Contact list, but without the need for account numbers or passwords (unless this information is necessary).

Software and Applications

List all software applications utilized by the unit or team that would be required in a recovery situation. List the name and version number of the application or operating system, the number of licenses required, and at what point during the recovery it is required (in the first 24 hours, by day 2, by the end of week 2, for example). Record the location of backup copies (source). Be sure to list the platform it requires (PC, Sun, Mainframe) (Table 15.4).

Equipment

List the critical equipment your unit will need to function after the disaster. This should include, but is not limited to, desktop or personal computers (if possible, state configuration or hardware requirements, that is, amount of random access memory [RAM], hard disk, processor speed, CD-ROM drives, special attachments), terminals, printers, fax machines, calculators, phones, modems, tables, chairs, soldering irons, microscopes, shovels, and like equipment. Include model numbers and enter the quantity required for a short-term outage (usually defined as less than 5 days) and also for a long-term outage. Estimate lead time for the replacement of equipment (Table 15.5).

Table 15.4   Sample Software Resource Questionnaire Form

Function:						
Software Description	Version Number	Serial Number	No. of Licenses	Backup Location	Time Required	Platform

Table 15.5   Sample Equipment Resource Questionnaire Form

Function:

Equipment Description	Make or Model	Configuration or Other Spec.	Short-Term Requirement	Long-Term Requirement	Lead Time

Forms and Supplies

List all forms and supplies that would be required to conduct business for both a short-term and long-term outage (for instance, letterheads, vouchers, diskettes, and special forms). Regular office supplies, such as paper, pens, or staplers, do not have to be included because they should be readily available at an alternate location unless the amount is extraordinary. Include such supplies as food, glassware, shipping cartons, and raw materials. Indicate where these supplies are stored or from where they can be obtained (Table 15.6).

Vital Records

List the critical records, reports, or documents on which your unit depends—records that must be available to send to regulators or that are necessary to conduct normal operations. Examples include accounts receivable, floor plans, corporate minutes, technical manuals, and the disaster recovery plan (Table 15.7).

Indicate where these records are stored or located within the company and on what type of media (paper, microfiche, or computer tape, etc.). If the record is in off-site storage, or at a law firm, identify the company and its location on this form and in the plan.

Data Analysis

The information gained from the interviews and questionnaires is analyzed according to the scope and goals of the project. Data are combined from all business functions, to allow the planner and management to decide which are critical to the continued operation

Table 15.6   Sample Forms and Supplies Resource Questionnaire Form

Function:

Description	Make, Model or Lot Number	Location or Source	Short-Term Requirement	Long-Term Requirement	Lead Time	Quantity on Hand

Table 15.7 Sample Vital Records Resource Questionnaire Form

Function:				
Description	Media Type	Internal Location	External Location	Form or Other Number

of the organization and which are dependent on others. Their outage tolerances are determined, and recovery priorities for both the individual business functions and support systems, such as computer applications, are assigned. When drawing these conclusions, the planner must keep the following guidelines in mind.

Outage tolerances (RTOs) and critical functions are not determined solely on the numerical data. For one thing, their designations may change. Also, whereas the analysis may indicate that product A generates the most income, it may also carry the highest recovery cost; management may decide to recover product B first because it has the lowest recovery cost, or is a product they wish to emphasize at the time. In most cases, it is readily apparent from the data which functions are the most important, but the planner must obtain early verification of this from the sponsor or senior management.

Again, account for impacts only once each; be careful of duplicates. In a large or complex organization, it is easy to add the loss of a dependent function more than once when calculating the overall impact.

Do not deduct insurance coverage or expected claims reimbursement from the loss figures. Although it is possible to predict (assume) the maximum amount of reimbursement for claims, it is more difficult to predict when the reimbursements will actually be paid or how the total amount will be distributed over the life of the claim. During a recovery situation, claims are usually not filed all at once—the extent of the loss, the amount of documentation, the varying procedures between insurance companies, and the added workload on the claims departments during an area-wide disaster will delay or extend the process. Small and medium-sized companies have little tolerance of an interruption in cash flow caused by a disruption; for them, anything less than a rapid reimbursement may have fatal consequences. Figures adjusted for expected reimbursement may not adequately warn of this danger.

After the data from the interviews and the questionnaires are analyzed, verify the results with business unit management and with the CFO. This verification is important for keeping the data credible throughout the process.

Establish RTOs both within the critical processing time windows and during "normal process times"—that is, not end of month or end of quarter. List both in the recovery plan.

Presentation of the Data

Because the BIA is often used to gain management support for the continuity planning program or to justify program budgets, their results are presented to senior management. This presentation could be the most important step in the impact analysis. If it is successful, the planner will gain the political and financial support needed to complete the project. If not, the planner will have wasted many hours of valuable time and will face needless delay and frustration and may produce a plan that is less than effective.

The analysis presented to management must be credible. As stated earlier, a top-down approach will result in the greatest degree of success. If this approach is used, the planner will have few problems with the acceptance of the analysis. When presenting the results, ensure that figures are accurate and not misleading and that the data and the presentation are short and simple. Matrices, tables, and statistics tend to bore and confuse people, even at the senior management level. Slides which contain spectacular pictures of disasters and natural phenomena for background tend either to distract or cause the graphics to be remembered only for their scenery. Relationships and financial data are better represented graphically, either by pie, bar, or other types of charts. Whenever possible don't present expected occurrences in terms of probability—state them as a fact. For example, if the local Office of Emergency Services or the U.S. Geological Survey predicts that the probability of a major earthquake is 82 percent in the next 10 years, report, "I believe there will be a major earthquake during the life of our strategic plan" (assuming it is a 10-year plan).

Distribute a written report that includes supporting data, the impacts for the individual functions or units, and the combined impacts for the organization. Unless it is important to the understanding of interdependencies or other relationships to mention others, discuss only the impacts of the major functions and the impact to the organization as a whole. During the presentation, outline why you are there and what you expect the group to decide or to do as a result of the analysis.

Reanalysis

Finally, BIA should not be a "one-shot deal." The information in recovery plans must be updated regularly and the basic strategic framework of the plan reviewed annually. It is logical that the impact to the organization will change as the structure and strategic direction of the business changes. The impact analysis will help to identify these changes.

REVIEW

A BIA identifies the financial and subjective loss of business functions over time. Some definitions say it identifies the loss of "critical" business functions, but it is the BIA that determines whether they are critical or not. A major objective of the BIA is to determine what the outage tolerance or RTO is of the functions and business processes.

A BIA differs from a risk analysis in important ways. At this point, we are no longer concerned with the cause, probability, or effect of an outage. The BIA has no focus on hazard identification, at least from the standpoint of pure risk. The outage tolerance or recovery time objective is exclusive of its cause. Many believe the BIA is the most critical part of the continuity process because it forms the basis for future decisions and develops or destroys momentum and support for the project. Various methods exist to accomplish its objectives.

Data are best gathered through a combination of questionnaires and interviews. The BIA, through the questionnaires and interviews, is also used to obtain resource information to be included later in the business continuity plan. It is important that data and conclusions are valid and accepted and that they don't duplicate financial and subjective loss from dependent functions.

The findings from the BIA are included in a report and presentation to management, avoiding the use of distracting graphics and facts and figures, opting for graphical representation of the pertinent information.

It is important to update the BIA on an annual or other significant basis.

16 Plan Documentation

Eugene Tucker, CPP, CFE, CBCP
Contributing author

"The planning process delivers value to the organization by forcing its members to identify critical functions, place a value on the loss of these functions, and to think through the most effective recovery strategies. The importance of documenting these efforts is fundamental to the Business Continuity process. The lack of a plan will add delays to the organization's fight to recover, cause the misuse of scarce resources, and may ultimately set the stage for the failure of the business."
—Eugene Tucker, CPP, CFE, CBCP

A business continuity plan (BCP) completes the definition of continuity planning by scripting the instructions to continuity team members so they may implement strategies and list the people and resources necessary for the team to accomplish their tasks.

REQUIRED ELEMENTS OF THE PLAN

Business continuity planning can take many forms, depending on the applicability of regulations, the organization's culture and document control procedures, and its level of sophistication (or limitations) in the use of technology (such as having interactive plans on CD-ROM or web-based plans). Absent regulatory issues, the best format to use is one that the planner is comfortable with or the organization is used to and that satisfies the following conditions:

- The plan must be organized in a logical sequence.
- It must be "clean" and easy to follow. (Information that is difficult to find under stressful circumstances is useless.)
- It must be complete but not overly detailed. (It must contain sufficient information to allow someone who did not participate in the planning

process to understand what is required to recover the business. It should not contain information irrelevant to the task at hand, such as hazard identification, mitigation recommendations, or justifications for the program.)
- It must contain a glossary defining any terms used.
- It must assign responsibility for planning to individuals within the organization and describe the emergency lines of authority.
- It must outline specific resources and tasks required to carry out recovery operations.
- It must be flexible enough to address unforeseen events. (No two disasters are exactly the same; neither are their response and recovery. The plan must allow for midcourse correction and adaptation.)
- It must contain references to other plans or documents. (Although all the information necessary to the recovery effort must reside in the plan, the plan should not contain a restatement of detailed operating procedures or instructions, such as a multivolume operating system installation set. Necessary documents, references, and other plans that are not immediately required and cannot be included in the plan should be listed in the vital records section. These must be duplicated and stored off site for later retrieval.) Note: In electronic-based plans, these documents, if not too large, can be included as document attachments.
- It must state assumptions upon which the plan is based or by which it is constrained.

MULTIHAZARD FUNCTIONAL PLANNING

Multihazard functional planning is a format the Federal Emergency Management Agency (FEMA) suggests that governmental agencies use to develop their emergency operations plans. An effective BCP follows the multihazard functional format, especially if a recovery team methodology is used. More recently known as "all-hazard emergency operations planning," it is based on the premise that although the causes of emergencies and disasters vary, almost three fourths of them produce common response requirements. The jurisdictions can then develop task-based plans around these requirements or functions rather than around each anticipated hazard. Some hazards do produce unique needs; these needs, requirements, and responses are appended to the "basic plan," under this methodology.

The main part of the plan—known as the *basic plan*—outlines the overall emergency organization and its policies, assumptions, activation, and lines of authority. Its primary audience is the jurisdiction's executive- and management-level staff.

Functional annexes are sub-plans that focus on specific functions the jurisdiction will perform in response to the disaster. Shelter management, evacuation, and search and rescue are examples. Each annex may contain its own appendix of tasks and requirements for dealing with the specifics of a particular hazard. Standard operating procedures (SOPs) and checklists may be included in the annex.

CONTENTS OF THE BASIC PLAN

The basic plan should contain the following elements or sections[1]:

- Introductory material:

    - Promulgation document (outlines the authority and responsibility for the plan; usually signed by the chief executive officer [CEO])
    - Signature page (demonstrates that all response organizations have participated in its development and are committed to its success)
    - Dated title and revisions page
    - Distribution list
    - Table of contents

- Purpose statement
- Situation and assumptions, including the scope of the response and the assumptions upon which the plan and response are based. (*Situation* refers to a description of the justification or necessity for the plan.)
- Concept of operations, describing the overall strategy or approach to the response, activation of the plan, and coordination with other agencies
- Organization and assignment of responsibilities, delineating the emergency organization and reporting hierarchy. Each position is listed, along with an overview of its duties and responsibilities.
- Administration and logistics, defining the general administrative policies, such as financial and purchasing controls and procedures, resource management, and mutual-aid agreements
- Plan development and maintenance, describing the overall approach to the plan—in contrast to the approach to the response above. It explains responsibilities of the planners and the planning process and includes the requirements for plan maintenance and exercising.
- Authorities and references, highlighting enabling laws and the legal basis for emergency operations and referring to other documents with relevant information

PLAN ORGANIZATION AND STRUCTURE

Like the multihazard functional plan, the BCP consists of a basic plan and departmental or team plans (similar to annexes). The basic plan contains administrative and descriptive

---

[1]Federal Emergency Management Agency, *Guide for All-Hazard Emergency Operations Planning*, SLG101, 1996-723-006/83312 (Washington, D.C.: U.S. Government Printing Office, September 1996).

details required to implement the plan, as well as information that is common to the recovery effort of two or more (or all) teams.

The basic plan may contain the following sections:

- Table of contents
- Policy
- Scope
- Objectives
- Assumptions
- Activation procedures and authority
- Emergency telephone numbers
- Alternate locations and allocations
- Recovery priorities or recovery time objectives (RTOs)
- Plan distribution
- Training
- Exercising
- Plan maintenance
- Appendix
- Team recovery plans

Table of Contents

The plan should contain a table of contents after its title page. If the plan is electronic, it is useful to include document links to each subject heading.

Policy

The plan can briefly outline or reference the existence of management's policy to develop, exercise, and maintain the plan. Ideally, it will include a full policy statement, signed by the CEO. Some planners include a brief description of the responsibilities of separate divisions, sites, or departments for developing their own plans. The policy should mandate the planning process, including important elements like an annual business impact analysis, exercises, and maintenance. It must assign responsibilities and allocate budget. It should tie BCP performance to performance goals and objectives and bonuses.

Scope

The dimension of the recovery process encompassed by the plan is briefly but completely discussed. If the plan pertains to a single building or site, refer to it by building numbers, site name, and exact street address. Inform the reader of any pertinent functions, locations, or contingencies not included in the plan. If the plan includes crisis management and response issues, bring them to the attention of the reader.

Objectives

What, in general terms, will the plan accomplish? Objectives can include the following:

- Ensuring the safety of employees and assets
- Minimizing economic losses resulting from interruptions to business functions
- Providing a plan of action for an orderly recovery of business operations

Assumptions

List any assumptions upon which the plan is based or by which it is limited. It is nearly impossible to plan for the absolute worst-case scenario, in which everything ceases to exist. Most of us now know that California will not fall into the ocean (or worse, see all its residents move east). Common assumptions include the following:

- Buildings will be either partially or totally damaged or inaccessible.
- Most key personnel identified in the plan are available following a disaster.
- Alternate facilities identified in the plan are available for use in a disaster.
- Backup data and valuable papers located in off-site storage will be readily available.
- Critical resources will be available.
- Most employees are trained in facility evacuation procedures.

Including the above assumptions in the plan does not mean that strategies to mitigate their occurrence are not designed and implemented.

Activation Procedures and Authority

List the people or circumstances who have the authority to activate the entire plan, individual team plans, or multiple team plans. Typically, any member of the management team or the recovery coordinator can activate the entire or any portion of the plan, whereas team leaders can independently activate their individual teams, with notification to the recovery coordinator.

Some plans allow for a graduated activation (level I, II, III); some, in an effort to get to the head of the line when competing for space at a hot site, will let the type of disaster dictate how the plan is activated. For example, in a localized disaster, such as a fire in the data center, the team will first assess damage before declaring a disaster. However, if the disaster is regional, such as an earthquake, and the data center is damaged, the Activation Authority will immediately declare a disaster to the hot site, pay the declaration fee, and then assess the damage and only then decide whether there is a need to relocate to the hot site.

Disaster levels can be defined as follows:

Level I disaster. A level I disaster is one resulting in facility inaccessibility or loss of power or other critical services for an expected period of up to 48 hours (this time factor can change according to the recovery time objective). Damage from a level I disaster is not large in scale. It may consist of minor damage to one or more buildings, lack of access due to weather or city infrastructure conditions, or hardware and software problems.

Level II disaster. A level II disaster exists when the outage is expected to last 2 to 5 business days. Damage from a level II disaster is more serious than level I damage and may result in heavier losses to equipment and documentation (files, reports, contracts) due to a prolonged event, such as a fire or flooding.

Level III disaster. A level III disaster is one in which the disaster results in outage anticipated to last in excess of 5 days. A level III disaster is severe and could include the total destruction of one or more buildings, or service within the buildings, requiring significant facility restoration or replacement.

Data-centric or information systems team plans may use a different activation sequence, each of which may involve a specific response by a different group of people. If a disruption is expected to last less than two hours, a "stage 1" situation is declared, the team leader or shift supervisor is notified, and instructions are implemented by the on-duty personnel. If the outage is not resolved within this time frame, or if it is apparent that more time is required, the response will escalate to stage 2 or 3. This will invoke an additional set of instructions, notifications, or full recovery team activation. Organizations must develop activation sequences that make sense for their situation.

Emergency Telephone Numbers

Include in this section emergency telephone numbers that may be common to all teams. This will normally include police, fire, and ambulance (paramedics). List both the emergency number and a secondary number that connects directly to their dispatch center. Radio and television station numbers and call signs, poison control, local medical clinics and hospitals, utilities and transportation contact numbers can also reside here.

Alternate Locations and Allocations

Virtually all recovery plans include an alternate location to which the team can transfer its operations to continue critical functions and resume the delivery of service. Common alternate locations include the following:

- Hot or cold site
- Vacant or shared space in another portion of the building, site, or corporation
- Hotel conference rooms

- Mobile trailers
- Third-party contract manufacturers
- Training centers
- Vendors and suppliers
- Competitors
- Home

Although the company may have an overall relocation strategy, certain teams may relocate to different locations. All alternative locations should be listed in this section.

Allocation refers to the number of team members that will relocate. Not all employees are generally needed during the initial phases of a recovery. This number will indicate how many employees or team members are needed to arrange resources and space. Some planners include the total head count and the short-term and long-term space requirements (square footage) in this section, but this is best left in the facilities, real estate, or other team plan.

Hot-site or relocation floor plans to expedite the setup of workstations, communications, and systems should be located elsewhere..

Recovery Priorities or Recovery Time Objectives

The recovery priorities section is a listing, usually in tabular form, of the organization's departments, teams, or critical functions in order of importance as determined by the business impact analysis. The RTOs, as well as any times of the year (quarter, month, or other business cycle) during which the loss of these functions is especially critical, are also listed. This information is useful to the management team if a decision to reallocate resources becomes necessary.

Plan Distribution

List the plan recipients in this section. If the organization maintains document control procedures, they should apply to the BCP as well. Initial publication and revision dates are important to avoid confusion and to help ensure that plan recipients have the latest version. Large organizations will not list in the plan every person who is entitled to receive a copy. Senior members may be listed by name, the remainder by title or generic description such as "team leaders" and "team members."

As mentioned in other locations of this publication, the entire plan is not distributed to every continuity team member. This is appropriate only for the continuity coordinator, plan administrator, off-site storage, and the Emergency Operations Center. Of course, there may be others who need the "phone book" version of the plan, but team members must receive only the information they will need in a continuity or recovery situation. This will help them to remain focused in their duties, locate important information in a stressful situation, and maintain security of the information contained in the plan. This is accomplished by distributing only a copy of the basic plan and the individual team plan to the team members. Some planners argue that the team members should receive only their team plan, excluding the basic plan.

Paper plans are useful and necessary—when all else fails, the ultimate BCP tool, the pencil and eraser, becomes the base element. Paper plans can become cumbersome, especially because as updates are issued, the outdated plan should be collected and destroyed. In today's electronic age, plans can (and should) be converted to some type of portable data file for use on a laptop computer, PDA, cell phone, or Internet or intranet. Many of the BCP development software programs have the capability to publish individual plans on the World Wide Web or intranet, or can convert their output to a Word file, Adobe Acrobat file, or other type of format. Large organizations will control plan distribution, especially if an enterprise-wide database or BCP software program exists, through team leaders. The possession of the latest version of the plan can become an audit item.

Training

The training and orientation requirements should be briefly referenced in the plan.

Exercising

The exercise program requirements should also be referenced in the plan and should include generic schedules such as "all Tier 1 functions (functions with an RTO of less than 24 hours) will conduct at least one tabletop exercise and one full-scale exercise per year." Exercise scenarios and other related documents should reside elsewhere.

Plan Maintenance

Like training and exercising, the people (generally) responsible for maintaining the plans are identified, along with the method and frequency of updates.

Appendix

The appendix will contain the glossary, forms common to most teams, supporting information, and common documents. Team continuity plans can also contain their own appendices.

Team Recovery Plans

BCPs, especially those for large organizations, can be many pages long, rivaling the size of most large metropolitan area phone books. The mere size of these plans will cause most people to place them forever on their bookshelves, to be read only by an auditor. Access to all information in the plan is required, however, only by the continuity coordinator, possibly the management information systems director, and just a few others. It is best to distribute to each individual recovery team only the portion of the plan that is meaningful to it.

The recovery team plan will contain at least the following information:

- Review of pertinent information from the basic plan
- Overview of the team plan
- Team member contact list
- Scripted continuity instructions
- Resource listings
- Blank forms, contracts, and other documents

Information from the Basic Plan.   Activation procedures, communications, organization, and structure are summarized in the team plan. The duties and responsibilities of the team leader and how the team relates to the recovery are discussed. This will allow the team plans to function independently of the basic plan, keeping the information focused and the size of the plan manageable.

Overview of the Team Plan.   A review of the general business functions of the department or team, along with its critical functions and continuity or relocation strategies, are listed here. Briefly include other relevant information.

Team Member Contact List.   A listing of the team members is given, detailing their titles or positions and emergency contact numbers (work, home, cellular, pager, and home email address). Call trees, if used, are placed in this section. It is possible that all team members will not be needed during the early stages of the recovery. Indicate at what phase or day they should be called. New regulations in some countries severely restrict the publication of home addresses and telephone numbers. Restricting this information to only team members may help, but as much contact information as possible should be included.

Many companies establish a toll-free number and team voicemail boxes at off-site (distant) locations for team members to call to receive information or instructions. The ability to notify and communicate with team members (and with the Emergency Operations Center) should have as much divergence as possible.

Scripted Continuity Instructions.   Include the instructions or tasks the team must follow to implement its continuity (recovery) strategies. Separate instructions can exist for the team leader and team members; they can be further separated by time (initial 24 hours, initial 48 hours, more than 5 days), by phase (response, restoration, recovery), or by whatever period is most practical. As stated before, these instructions should be brief but complete. They should be understood by someone reasonably familiar with the type of operation the team is attempting to recover; avoid, however, technical terms, abbreviations, and acronyms. The instructions should minimize the need for decision making after the disaster.

Resource Listings.   Vendors (internal and external contacts), customers, equipment requirements, forms, supplies, software and applications, vital records, and other resources are listed in this section. Model numbers, configurations, floor plans, cabling diagrams, tax tables, milestones, and other due dates, along with both the short-term and long-term quantities required, are listed.

With vendors and contractors, list as many contact numbers (home, office, pager, fax, and email) as possible. If there is a corporate headquarters, list this number also. Include purchase order numbers, license numbers, and passwords as necessary.

A complete customer list with contact numbers could be appended to the plan, or critical customers who need or should be notified of the outage can be included in the plan. Consider calling customers to inform them of a new temporary address or of the firm's continued ability to deliver its services. Consider completing these notifications even if the company is not directly affected by the disaster. Firms in San Francisco lost some business after a major Los Angeles earthquake more than 500 miles away because the customers did not understand the geographical separation or the nature of earthquakes.

Blank Forms.   Forms that are needed by the team and included in the appendix are also found here. Copies of contracts or agreements with vendors should appear in the team plan.

17 □□□
□□□
□□□

Crisis Management Planning for Kidnap, Ransom, and Extortion

"Kidnapping, abductions and hostage situations have been a human practice since the beginning of time. These practices, however, have changed radically in the last 15 years."
—Charles Seoane Norona, CPP, ASIS International Seminar on Kidnapping, Dallas, Texas

Many business executives, continuity planners, and security directors incorrectly believe the term *crisis management* is synonymous with *business continuity planning*. There are important distinctions between the two, although the methods used to arrive at plans for each are basically the same (see Chapter 16). Planners who miss these distinctions believe their plans will allow them to respond and recover from all major threats to the organization. This mistake can cost the firm millions of dollars in lost market position, diminished reputation, and devaluation of the company. Union Carbide illustrated this point after the value of their stock declined 27 percent immediately after the toxic gas leak in Bhopal, India. In 1989, Perrier was the market leader in bottled mineral water, with sales over 1 billion bottles per year. Their bottling plant in Vergezem, France was tooled for a projected market growth of 20 percent. One year later, the news media reported that benzene, a chemical linked to cancer, was found in the water. Sales plummeted to 500 million bottles, and market growth was essentially nil. The Vergezem plant could no longer operate profitably, resulting in its acquisition by Nestlé.

The security profession uses the term crisis management to mean terrorist acts, kidnapping, labor violence, civil disorders, industrial disasters, and natural catastrophes.[1] The insurance industry uses the term to mean the response to (but not the recovery from) an emergency or disaster such as a fire or tornado.[2] To further complicate

---

[1]John J. Fay, *Butterworth's Security Dictionary—Terms and Concepts* (Woburn, MA: Butterworth Heinemann, 1987).

[2]George L. Head, editor, *Essentials of Risk Control, Vol. II.* (Malvern, PA: Insurance Institute of America, 1989).

our understanding of this concept, some business continuity vendors use the term to mean a hot-site recovery operation.

Both a crisis and a disaster will affect the financial position of the company. The goals of their response are essentially the same, but continuity planning is generally focused inward toward mission critical business processes (functions), whereas crisis management is focused outward to manage the image or public perception caused by an incident or disaster. Crisis management planning ventures slightly outside the traditional role of the security professional by the inclusion of speculative risks in both the analysis of threats and the design and implementation of response and recovery strategies.

Maxialo Ckonjevic defines a crisis as, "Extreme threats to an organization which have the potential for significant negative results to important organizational values, functions and services and which could result in major damage to the organization, its employees, products, services and reputation."[3]

It is possible to experience a disaster without a crisis. The reverse is also true. The temporary loss of the Exxon Valdez after it spilled 250,000 barrels of oil into Prince Edward Sound was not initially a disaster. Oil wells continued to pump, transportation and billing systems functioned, and no refineries went off-line. But because of the manner in which the company responded to the incident, Exxon lost many uninsured dollars in goodwill and future sales and gained closer regulatory scrutiny. I know one person who would still push his car past an Exxon station if he ran out of gas because of Exxon's seemingly uncaring corporate image. A disaster can escalate to a crisis if a business does not recover within its outage tolerance (recovery time objective).

Business continuity focuses on the loss of critical business functions. Crisis management considers loss to its image and goodwill. Major components of crisis management include communication and public information plans. Management is most likely thinking along these terms when defining crisis management.

Crisis planning can conform to the comprehensive emergency management model. Once the threats are identified, preventative measures are developed and implemented. Examples of these measures include tamper-proof product containers, corporate procedures, and controls. Preparedness can take the form of predeveloped statements or press releases, the selection of a company spokesperson, and media training for the management team. Response includes the rapid collection and dissemination of information designed to minimize the negative impact of the incident or to prevent the escalation of a situation to a crisis. Recovery includes strategies to restore public, employee, or shareholder confidence.

This chapter examines the framework of crisis management planning and places it in the perspective of planning for kidnap, extortion, and ransom. The odds of any business organization being the victim of a kidnap, extortion, or ransom demand are slim. Nevertheless, I outline here the steps to be taken by a crisis management team (CMT) because a kidnapping could nonetheless occur at any time. More important, however, the CMT approach can be used to resolve any crisis of magnitude sufficient to threaten the financial existence of a corporation. I go into specific details regarding the threat of kidnap and extortion because I have worked with organizations that, not having plans, had to develop them on an ad hoc basis to deal with these problems. This real-world

---

[3]Max Ckonjevic, FBIC, CGCP, Presentation to the "Survive" conference, San Francisco, 1997.

experience is an example to the reader of how to develop, before it occurs and not after, a planned approach for dealing with such a crisis.

THREAT IDENTIFICATION

Some companies spend billions of dollars investing in their corporate image through trade and brand name identification. The value of this identification could represent the company's largest asset. Threats against it must be identified. Threats outlined in the crisis management plan include the following:

- Accusations against, or the arrest of, a company official
- Boycotts
- Demonstrations
- Errors and omissions
- Fines and penalties imposed by regulatory agencies
- Missed production milestones
- Good neighbor policies and community concern
- Hazmat and environmental issues
- Health of key executives
- Hostile takeover attempts
- Human error
- Insider trading
- Kidnap and ransom
- Labor disputes
- Major injuries
- Mismanagement
- Negative research or a death from clinical trials
- Patent infringement
- Product liability
- Product recalls and defects
- Product tampering
- Reorganization
- Rumors
- Terrorism
- Third-party crime (assault, rape, robbery)
- White collar crime
- Workplace violence

These risks are best identified through a risk analysis and crisis scenario planning than with a business impact analysis. Use scenarios to uncover the sources of a crisis. When developing scenarios, look at situations the competition has experienced or is likely to face. Select each anticipated event and subject it to as many "what if" questions as practical, such as, What if an executive is kidnapped domestically, overseas, for ransom, by what group? and so forth. See Chapter 12, Mitigation and Preparedness for a brief explanation of scenario planning.

Government officials and business executives have been attractive targets for kidnappers and extortionists since long before the 1960s. Domestic terrorist activity directed against executives, corporations, and political leaders in Asia, Europe, South America, and recently Mexico and the United States has made the general populace justifiably apprehensive. As in the case of the Hearst family, corporations, their executives, and members of the executives' families have been, and are expected to continue to be, victims of kidnappings and extortion, which could inflict heavy losses on corporations or on the personal well-being and wealth of the victim's family.

There are two basic reasons for executive kidnappings—personal gain and political objectives. These kidnap (extortion) victims generally possess one or more of the common elements of money, power, or high public visibility. Individual chances of being kidnapped are extremely low, but the odds increase rapidly if the potential victim is wealthy; controls large amounts of money; is associated with such cash-driven industries as banking, savings and loans, gambling casinos, or food marketing; or works in industries such as airlines or public utilities. An executive's chances of being kidnapped are further increased if his or her industry is often victimized by terrorists or extortionists, or if his or her company has a history of paying ransom demands.

Plan Documentation

To mount a successful response to kidnappings, extortion, and other threats, a plan to deal with such crises must be formulated in advance. The major responsibility for advance planning belongs to the organization. Organizational planning is more effective than individual effort, and it is more likely to be implemented and thus successful. Therefore, organizations must develop crisis management skills that are adaptable to any demand made on them.

Although a crisis management plan can stand on its own, a business continuity plan that does not include crisis management will generally fall short of the objectives of the organization. Crisis management can exist as a subsection of the business continuity management team plan, as part of the public relations team plan, or as a combination of both, depending on the construction of the document and the size of the organization. Some planners believe the development of separate documents for continuity planning and crisis management is the most effective approach. Others point out that separate plans are cumbersome and require extensive duplication of information and upkeep. The complexity of the risks will guide the planner toward the best document format.

Because extortionists can often inflict heavy losses on organizations, it is imperative that the CMT prepare a readiness plan that will minimize these losses. This plan must fix corporate objectives and limitations, and it must be designed to be effective when the CMT is operating under the emotional strain of responsibility for human life, often with limited data and time for making decisions.

The plan must resolve the fixed elements of a crisis, so as to require the CMT to make only those decisions during a crisis that are affected by immediate variables. Also, it must have sufficient flexibility to enable the CMT to develop alternative strategies after gathering information and analyzing threats under rapidly changing crisis conditions.

In the event of a kidnapping, provisions for gathering personal data, such as employee and family biographical sketches, as well as medical and other requirements of the employee and his or her family, must be incorporated into the plan along with methods to make these data readily available during the crisis period.

The resource section of the plan should include phone numbers and addresses of the team members, major customers, media contacts, brokers, local officials, and regulatory agencies. The plan should include instructions on how to best contact these resources and officials. Ensure that equipment and supplies such as projection machines, sound amplification equipment, battery-driven bullhorns, sign-in sheets, and name tags are available.

The crisis management plan, like the continuity plan, is a confidential document. It contains important strategic information and phone numbers of key executives. Its distribution must be limited. Both are living documents that must be maintained and simulated. Testing of the plan should include role playing and on-camera interviews. Rehearse responses well in advance of the need.

PLAN ACTIVATION

Activation of the crisis management plan is a bit more difficult than the continuity plan. An earthquake, fire, or system crash is an obvious sign of trouble, but the signals of an impending crisis are often less obvious. Subtle signals can appear in trade journals, a rise in terrorist threats, or an increase in customer service calls. It is important to distinguish among an issue (a matter of dispute and differing points of view), an incident (a sudden, unexpected occurrence that requires prompt action), and a crisis. A crisis is far more crucial than most issues and incidents—it is a major turning point resulting in permanent and drastic change to the organization. These events can have an internal or external impact (fire or earthquake), can be sudden or gradual (product sabotage or community pressure after a toxic release), can be short-term or long-term in nature (bomb threat or months-long strike), and can be human in origin (rumors about product safety). The failure to respond at the appropriate level can lead to serious errors, spread alarm, and create a siege mentality. The analysis of the threats should include a list of indicators or symptoms of a potential problem so that the team can evaluate and respond to them as quickly as possible. This response should be proactive; a nonreactive position may place the organization in a defensive position.

Institute a system to monitor emergencies, problems, and controversies. Clipping services that capture news reports about the company from radio, television, and newspapers are a good source of information that can warn of an impending crisis or provide feedback on how the company is perceived after a crisis. Summaries of complaints from customer service or technical support should be tracked for this purpose.

CRISIS MANAGEMENT TEAM

Crisis management can be addressed most advantageously by a CMT. The CMT should consist of a group of senior management personnel who have the authority to make decisions for the entire corporation during a crisis. Because a small unit is generally capable of reaching decisions more quickly, the CMT should consist of the least number of individuals possible. It should also be dynamic in nature and should consist of a team leader, a secretary, and a senior member from the following departments:

- Executive management (chief executive officer)
- Finance (chief financial officer)
- Operations (chief operations officer)
- Risk manager
- Sales and marketing
- Legal and/or regulatory affairs
- Public relations
- Human resources
- Telecommunications
- Security

On a CMT, the chief executive officer can act as the team leader. A secretary or administrative assistant is included to log events, actions, times, and expenditures. As with continuity planning, a complete and accurate record of events will help the company wade through the inevitable legal, regulatory, and insurance problems once the crisis is stabilized. Technical advisors can substitute for the chief operations officer to assist the team with the development of consistent and accurate assessments, projections, and press releases. Technical advisors can include hazardous materials experts, the Federal Bureau of Investigation, and rescue or hostage negotiation teams. A representative from the telecommunications department is included to help the team decide if any extra load on the telephone switchboard can be accommodated by the installation of extra trunk lines, embargoing nonemergency calls, and arranging for additional switchboard operators.

The CMT will collect and evaluate information on the scope of the incident and attempt to prevent its escalation to a crisis. Team members will review strategies designed to mitigate the adverse effects of the incident or crisis and attempt to benefit from any opportunity the situation may present. The repurchase of stock or the suspension of trading, increased advertising, or public service announcements (or the suspension of advertising) and the recall of products are some strategies the team may select. They will look at how organizational dynamics, financial position, and public opinion have changed since any strategies in the plan were first developed.

The CMT is not a substitute for law enforcement or the organization's security department; rather, it is a complementary support organization. Decisions that affect the company directly and require corporate decisions or responses should be handled by the CMT in partnership with law enforcement. For example, a trained hostage nego-tiator working in conjunction with law enforcement personnel should be designated to act as the intermediary between the extortionist and the victim company. The extortionist must clearly understand that the negotiator has neither the authority nor the capacity to make decisions or commitments on behalf of the company. Used in the proper context, the trained negotiator often provides time for the CMT decision process to work effectively.

Most firms have capable, qualified, and responsible people in their employ who can and should conduct all negotiations in these situations. These people, many of whom are schooled and experienced in industrial relations or similar types of business negotiations, need only receive training in how to negotiate in criminal situations in order to become effective.

The use of outside consultant resources for negotiations is impractical, for several reasons. Most extortion situations are resolved rather quickly. The one exception is a protracted terrorist kidnap, for which publicity is a key ingredient of the case. By the time outside consultants could arrive at the scene and receive an update on the progress of the case, local law enforcement, combined with company executives, would probably have the situation stabilized and in some cases resolved. Coming on the scene 6 to 12 hours after a kidnap-extortion situation begins, the outside consultant will likely have no impact whatsoever. In addition, remember that the basic tenet of risk control consulting is to make the client as self-sufficient as possible in matters pertaining to security; this includes crisis management programs. The best use of consultants is to have them help the corporation to develop the CMT, then to have them available for telephone consultation during the crisis.

There are several basic areas of concern to be addressed by the CMT as it provides corporate leadership during a crisis. The protection of assets, which in this case includes personnel, is of primary concern. Experience with a systems approach to assets protection, as well as a knowledge of the types of adversaries encountered, should be provided to the CMT by the organization's security director (consultant), who can also assist by acting as liaison with law enforcement agencies.

The CMT will require the assistance of its legal counsel to examine such issues as employee and stockholder rights in relation to the legal standing of the company regarding various strategies and monetary payments to extortionists. Information from the financial arm of the corporation is needed to develop the monetary base for CMT operations, and its assistance is needed to set the corporate strategy and limitations regarding ransom of any particular corporate employee.

The CMT must also consider the long-term effect of crisis decisions on employees of the company.

The CMT must be given autonomous control over decisions the corporation must make during a crisis, consistent with an advance plan approved by the board of directors. Every action to be taken by the company not dependent on the specific nature of the crisis should be rehearsed, much like a fire drill; only the variable decisions will then have to be handled. Even those decisions will be addressed from a perspective of preset goals, limitations, and strategies.

A corporate crisis-management capability will enable professional law enforcement personnel to respond to the crisis with better initial information and a clearcut base from which to operate. Also, this capability will lessen considerably the probability of loss through matters growing out of the original crisis, such as stockholders suits, employee negligence suits, wrongful-death suits, insurance cancellations, and expropriations of assets by foreign governments irritated by the way in which a corporation handled a problem.

HANDLING THE INITIAL CONTACT

When an extortionate demand is received, the CMT, the organization's security department, and law enforcement should be advised immediately and the crisis management program put into action. The actions taken during the first crucial moments after an extortionate demand is received may well determine the eventual outcome.

Because most threats are transmitted by telephone, recording devices and tracing capabilities should be discussed with the local telephone company. Recording an extortionate call will not only preserve its details for later analysis in decision making but also may provide investigators with background noise and voice-print characteristics, leading to the place of origin of the call and the identity of the extortionist.

People who may receive initial extortionate communications are, in many respects, vital sources of information for the CMT. As such, a training program should be developed to ensure proper implementation of the procedure regarding handling this type of call. Individuals handling such calls should be instructed to remain calm, record or write down all data given by the extortionist, express cooperation, and ask questions to lengthen the time of the call. An attempt should be made to calm the extortionist and secure proof that the hostage is being held and is unharmed. The recipient of the call should attempt to talk with the hostage and to give the hostage the opportunity to relay critical information through a prearranged code. Above all, people receiving the initial call should bargain for time; if possible, they should end the conversation in such a manner that additional contacts with the extortionist will be necessary before a ransom is paid. This allows for the opportunity to trace and record a second call from the extortionist, as well as providing time to implement the crisis management program and setting the stage for a controlled negotiation response.

Many extortion demands are transmitted in the form of a written threat. The letter and its envelope should be protected from unnecessary handling and preserved for fingerprints, handwriting, and printing and typewriting examinations, as appropriate. Following receipt of a written threat, steps should be taken to identify the source of the document if it was not mailed. It may be necessary to interview all employees immediately in order to develop information leading to the identity of the person who delivered the communication.

During the initial phase of the crisis, it is imperative to determine whether the demand is a hoax. In a kidnap case, the whereabouts of the alleged victim must be established immediately. Employee family biographical fact sheets can be of critical importance at this time. Several notable kidnap hoaxes have involved calls to the executive's family in which the caller pretended to be a telephone company representative. The caller would state that the family telephone was being serviced and request cooperation in not answering the telephone for the next hour. The executive would then be called at work and told his family has been kidnapped. Naturally, when he called home and got no answer, he would panic and comply with the extortion demands, believing that his family has been taken hostage. A family fact sheet containing the telephone number of friends and neighbors who can confirm the whereabouts of the family can be one means of thwarting such a scheme.

RANSOM CONSIDERATIONS

Payment of ransom is a decision to be made solely by the corporation or the victim's family. Law enforcement officials will discuss the pros and cons of ransom payment with the top officials of the organization and with the family of the victim. They will not, however, make the final decision as to paying or not.

Policy and the limitations on payment of ransom should be developed by the corporation and approved by the board of directors. This way, directors or executives insulate themselves from civil liability. For example, a shareholder could allege that the executive approving or making the ransom payment had not acted legally, did not have corporate authorization, and therefore was personally liable to the corporation for the amount diverted. If the payment of ransom or any other action taken in response to the extortion demands was itself a violation of local criminal law, the civil liability position could be aggravated. Finally, if the executive approving or making the payment failed to consult other executives or directors but was nonetheless able to obtain the cash or other assets and complete the transaction, shareholders could allege that the other executives and the directors were negligent in failing to consider the possibility of such extortion and in failing to require appropriate controls. In such a case, liability might be alleged against all involved.

Inadequate action or improper action by the corporation leading to death or injury of an employee might result in claims against the corporation for damages by the employee or his family. This is particularly true if there were no contingency plans and the injured employee was exposed chiefly because of his corporate employment.

The above examples are not all-inclusive, and corporate counsel should be consulted in all such matters. This is not to suggest that no action be taken to free a kidnap victim merely because of the potential of civil liability. Instead, it emphasizes that the way to minimize or avoid such liability involves preplanning and prior authority.

It is not possible to fix any categorical limits on the amount that should be paid for the release of a kidnap victim. In most companies, the only likely financial gauge of the impact of the death of an executive or official is the amount of "key man" life insurance contracted by the company. This type of insurance is intended to cover the cost to the firm of replacing a deceased official and the interim expenses or losses likely to result from his or her sudden absence, and it does not address the sensitive area of public and employee attitude toward the company; nonetheless, the amount of such insurance is at least a rough standard that can be used as a first step in considering ransom payment amounts.

Another alternative in this situation is to refuse to pay ransom altogether. It has been suggested that paying a ransom of any type may induce others to try again, that the possible individual loss of life is a necessary cost. This is the position taken by many governments in regard to the kidnap and ransom of government officials, but it may be untenable when applied to private business enterprises. At least within the United States and Canada, the business community and the general public may not accept the position and its potential cost.

If a decision is made to pay a ransom, the net impact on the enterprise may ultimately be much smaller than the amount paid, for two reasons. First, active cooperation with law enforcement from the very beginning will improve the chances of capturing the kidnappers and recovering all or part of the money. Second, commercial insurance can be purchased to cover a portion of the ransom payment actually made.

If the decision has been made to meet ransom demands, law enforcement authorities will assist in preparing the ransom package. Plans must be made for availability of funds in appropriate denominations. It takes considerable time and effort to record currencies used in ransom payments, and this step should be completed as much in advance of the payoff as possible. Large amounts of money, in small denominations, produce surprisingly heavy, bulky packages; $1 million in $10 bills weighs about 225 pounds, for example. This should be kept in mind during negotiations.

Where nonmonetary demands are made, such as supplies, publicity, or chartered aircraft, the responsibility must be the result of thoughtful decision on the part of the CMT. In formulating a policy, the possibilities for ransom should not be limited to money. However, political demands, such as the release of prisoners and the provision of arms, generally cannot be influenced by officials of the kidnapped victim's enterprise.

PREVENTIVE SECURITY

Unfortunately, experience has shown that a dedicated group of terrorists (extortionists) can penetrate all but the most sophisticated personal and corporate security systems. However, a company's demonstrated crisis response capability and awareness by executives of personal and corporate security practices will likely decrease the chances that a corporation or its executives will become victims of a kidnap or extortion attempt. The key is to alert the corporation and its executives to the level of danger where the particular executives reside, the need to avoid patterns or routines in personal behavior, and the increased vulnerability of wives and children. These are all important factors in forming an appropriate preventive security plan.

Some of the more obvious suggestions include steps toward ensuring the physical security of executive residences, instructing children on appropriate precautions, limiting the dissemination of personal information to only those deemed in need of such information, and securing automobiles. Additional and more detailed security precautions can be developed through consultation and internal planning.

SUGGESTIONS FOR KIDNAPPED INDIVIDUALS

Based on information developed in past cases, it is clear that kidnapped individuals should control their fear and realize that professionals are working for their safe release. Problems should be analyzed and decisions made based on the individual's present condition; a display of anxiety could be contagious and counterproductive. In many instances, a victim can actually control an abductor's actions, through his or her dominating personality, leadership qualities, and calm approach to the situation.

If the victim is troublesome or appears to jeopardize the abductors' plan, serious harm to the victim may result. The victim should attempt to convince his captors that his or her well-being is essential to their success. Those working for release will be simultaneously making every effort to convince the abductors that their goals will not succeed under any circumstances unless the victim is set free, alive and unharmed. An attempt should also be made to develop a relationship with the abductors so as to change their perception of the victim from that of an "object" to that of a "person," similar to them. Attempts to cooperate with the abductors should be made with this in mind.

If given a chance to communicate with people working for their release, victims should attempt to give maximum information through prearranged code words, phrases, or verbal mannerisms that have been developed by the CMT. If the victim recognizes the captors or any detail of the kidnapping, it is imperative that this knowledge be kept from his captors because it may cost the victim his or her life.

There are almost no circumstances in which an escape attempt is recommended. The key word, however, is "almost." The possibility of escape should not be considered if the victim is goaded by impatience. Escape attempts should be viewed as a last resort, not a time-saving device. By considering the abduction a long-term venture, the victim will be less tempted by impatience. Escaped victims could find themselves lost in a remote, inaccessible, alien region, without transportation, money, food, water, or shelter, and perhaps unable to speak the local language. If recaptured (and not killed or seriously injured in the process), the victim will likely be treated more harshly. Thus, escape should only be considered as a life-saving effort when success is reasonably certain and the likely alternative is death.

MEDIA CONTROL

The CMT will also review methods to best communicate the position of the company to the public. The public may include the news media, stockholders, employees, customers, potential customers, regulators, and potential victims. The company must present a consistent, coordinated response. Establish media policy and communicate the policy to all employees. Employees should be instructed not to speculate on conditions and to refer all media inquiries to the appropriate spokesperson. The team will also decide what information not to release. Extortion or the kidnapping of an executive is best kept confidential.

The premature release of information could have an overwhelming and demoralizing effect on public or stockholder confidence. A premature "do not use" directive by a water department or a suspected food contamination warning will not only satisfy the goals of a terrorist but also can cause economic impacts that are well documented. A delayed disclosure could cause severe illness or injury, increase civil and regulatory liability, and diminish the credibility of the entity's members. The failure to release any information in some cases could be the best course of action (such as with a bomb threat) but could also have negative consequences that the firm should be prepared to address.

The most important resource for the CMT is accurate information on the scope of the problem. Effective crisis management often means effective information management from the standpoint of both rapid access to data and conditions for the dissemination of information to the public. The manner in which information is controlled, and who is in control of the information, will ultimately shape public perception. Use information to enhance the firm's credibility. Train the management and public relations team members (if your organization is large enough to have a separate public relations team) in on-camera media relations. Role play as much as possible as a part of this training.

A media relations expert should be available to the CMT for a positive, controlled response to media inquiries during the crisis—because one asset to be protected is the public image of the corporation. In this regard, it is imperative that responses to the media be coordinated with law enforcement officials to avoid premature release of information, which may jeopardize a victim's life.

The Emergency Operations Center or Command Center where the management team meets should be off limits and out of sight of the media. Any sign of stress or confusion should not be shown to the outside. Assign a person to direct visitors and relatives or friends of victims to a quiet room away from the press. Keep press, visitors, and victims apart.

Establish a secure room near the entrance for reporters and equip it with the latest technology used by the press. This can include extra phones, facsimile machines, data ports, satellite hookups, and so forth. Obtain equipment in advance to set up and support logistical needs for the largest number of reporters that can be expected. Assign a public relations representative to the press room to answer questions (within limits) and to make other decisions and arrangements. Establish written ground rules for press briefings. Arrange to have the facilities and telecommunications team set up live and taped information, and provide sign-in sheets and name tags with affiliations. Maintain a log of media contacts, and videotape all media briefings.

Establish compendia of information before an incident that include the following:

- Binders containing all recent press information, background on the company (its general description, employees, products, income, donations and charities, awards, inspection certificates), media advisories for the event, and special information phone numbers. Distribute these as necessary.
- Produce background ("B-roll") raw video footage about the company and its facilities and products. Make copies available to broadcast media in the event of a disaster. This will help the company control information, help reporters meet deadlines, and possibly prevent news crews from filming events or locations the company may wish not to broadcast.

Depending on the type of the crisis, consider establishing a public fact center, separate from the press office, to disseminate information to customers and the general public. Use this center for rumor control with live and taped telephone response. Update taped information on a separate line.

CONCLUSION

Business continuity and crisis management are similar in that they both involve preplanning that minimizes the need for decision making after the plan is activated. Because much confusion exists about what the term *crisis management* implies, the security manager is often responsible for developing this capability simply because management does not know where it belongs in the organization. The risk analysis process is used to identify and categorize the exposures. Operational and media strategies are developed, implemented, tested, and maintained through the use of scenario planning. A primary focus of the CMT is the identification of an impending situation and the control of public perception to prevent an incident from escalating to a crisis. A crisis management plan is developed and formatted in the same manner as a business continuity plan. The plan will allow for a rapid, proactive response. These plans can be combined or can exist as separate documents.

To prevent or minimize the harm that might result from executive kidnappings and other forms of extortion, the business community should recognize the need for, and take the necessary steps to develop, crisis management plans. The responsibility for developing such plans lies with the corporation itself. It is only through such planning, both internal and in consultation with experts, that the tragedies inherent in such crises may be avoided or minimized.

For more detailed information, see Appendix E, Sample Kidnap and Ransom Contingency Plan.

BIBLIOGRAPHY

Barton, Lawrence. *Crisis in Organizations: Managing and Communicating in the Heat of Chaos.* (Mason, OH: Southwestern, 1993).

Ckonjevic, Maxialo, FBIC, CGCP. Presentation to the "Survive" Conference, San Francisco, 1997.

18 Monitoring Safeguards

"In God we trust, everyone else we monitor."
—Anonymous FBI inspector

The security professional may be called upon to review or design all the security for a company. If a facility has a security system in place, the security professional may be asked to review it. The system, upon review, will prove to be either adequate or inadequate in meeting the needs or objectives of the company, which generally can be defined as the protection of assets.

MONITORING OR TESTING THE EXISTING SYSTEM

One technique for making an adequacy determination is to monitor or test the existing security program and systems to determine whether they are still doing the job for which they were initially designed or installed.

I was once assigned the task of evaluating the security system for a research and development (R&D) division of a large computer manufacturing firm. The R&D division, in existence at the same location for 5 years, had taken over facilities previously occupied by the consumer products division, which had been dissolved during a reorganization 5 years earlier.

All the administrative support systems, including security and safety, already at the facility had remained in place to serve the newly expanded R&D effort. In fact, the only significant change at the facility was that instead of manufacturing digital wristwatches and calculators, the facility was now developing artificial intelligence data and advanced computer-aided design information, for the U.S. Department of Defense (DOD). Five years later, however, at the time of our survey, the security system was found to be still protecting digital wristwatches and calculators!

We were asked simply to design a test to answer the question, "How well is the R&D facility protected against industrial espionage?" The answer was obvious even without testing: the R&D programs were totally unprotected! In 5 years of existence at this location, no one had bothered to review or to test the security program and systems to see whether they were still meeting the client's objectives. The client was spending in excess of $750,000 a year to protect products from theft and industrial espionage the company had stopped producing at this location 5 years earlier!

One might question why a multinational corporation with vast financial resources at its command would ever allow a situation such as this to exist. The answer, sad though it may be, is that such situations are not the exception; they are generally the rule. Situations like this are seen many times at many locations. Absent serious security problems, productivity is generally the most important issue and the one consideration upon which management's attention remains focused.

From the standpoint of achieving a functional security system, one that can be counted on to work when it is needed, testing of safeguards (countermeasures) in a production environment is probably one area most likely to be overlooked. This is in sharp contrast to the scientific and engineering fields, in which the need for periodic testing is usually accepted as an article of faith.

THE SCIENTIFIC METHOD

Most high school students have had an introduction to the "scientific method," the basis for all modern science and technology. The scientific method is, simply stated, a very basic problem-solving approach, namely, the gathering of data to be used to confirm or reject a developed hypothesis.

Few of us will object to the statement, "People and procedures must be tested in a number of scenarios, and testing, to be effective, must be an ongoing process." This same principle, however, is seldom applied in the real world of security. Yet, in no other way than by periodic, programmed testing can the integrity of any system or procedure be proved or, conversely, system flaws become detected, before the system fails and catastrophe strikes.

Depending on the type of security in question (procedures, hardware, electronics, or manpower), testing can take many forms and have many objectives. Here we are mostly concerned with tests that evaluate performance and reveal weakness, failures, or potential flaws in the design of the system—testing that will uncover problems that otherwise might remain undetected. Periodic tests, from a security perspective, are invaluable and should be included in every designed security program and system, no matter how large or small.

FIVE BASIC TYPES OF TESTING

There are five basic types of testing, which can be summarized as follows:

- *Functional testing:* determines whether hardware, such as a closed-circuit television camera or an electronics access control system, will do what it was designed to do
- *Safety testing:* determines whether the object or a procedure can be used without causing injury, loss, or harm
- *Performance testing:* normally concerned with conformance to timing, resource usage, or environmental constraints (an example is an anti-intrusion alarm)
- *Stress testing:* checks a person's or an object's tolerance to abuse or misuse under deliberately introduced stress techniques

- *Regression testing:* usually applies to an object, system, or procedure that has been altered to perform a new function and must still perform some of the functions for which it was originally designed

It is well to remember that testing can apply equally to people, systems, procedures, methodologies, and objects. Also, regardless of the application, testing must have a specified objective. From a security viewpoint, it is wise to question, before initiation, what the test objective is and why a particular test is deemed important. Other questions that should be answered include the following: "Are the tests adequate? Are the results valid? Can this type of test uncover a weakness or flaw that might otherwise remain hidden? Is this the best test to use?"

The best time to perform a hardware or electronics equipment test is during design, or at the latest, during installation, before acceptance. This facilitates changes, enhancements, and deletions. Efforts should be made to facilitate testing. Many electronic circuits are designed to include self-test circuits, diagnostic lights, fault detectors, and other built-in test aids.

Tests can be broken down into component segments. This makes it possible to test various sections at different times. It is sometimes more desirable to conduct functional tests this way. Performance testing is another area in which it may be desirable to test only parts or segments of the whole system. The more complex the system, the more difficult, time-consuming, and expensive it is to test the complete system at one time. The modular or segment approach can be designed to be reliable and less time-consuming, and from management's standpoint, to have less impact on production schedules.

Depending on the facility involved, tests can be relatively easy, such as verifying and updating biographical data. Tests can also range from the use of "tiger teams" and mock penetrations to that of a complex security environment, such as a DOD facility. Most tests fall somewhere between the simplest testing and complex penetration efforts, but each type is an essential part of the security program and should not be overlooked. The results of all tests should be studied thoroughly and subjected to interpretation by objective analysis.

AVOID PREDICTABLE FAILURE

It is also desirable to reduce one's identified risk by testing. Murphy's Law states, "Anything that can go wrong, will." A relevant corollary could be, "Any system that is not periodically tested will eventually fail." Systems that are routinely tested also occasionally fail! The idea, however, is to avoid predictable failure. The testing concept is consistent with the objective of reducing risk by diminishing uncertainty, which is, after all, one of the fundamental principles of an effective security program.

With all the obvious benefits to be gained by testing, one might logically inquire, "Why don't security professionals do more of it?" The answer is simple—time and money! Even routine testing is both time-consuming and costly. Testing complex systems, on the other hand, can be enormously expensive in terms of time and money. A routine fire evacuation drill in a manufacturing plant, for example, can cause thousands of hours lost to production.

SOME AUDIT GUIDELINES

In some cases, test expense can be reduced by use of some basic audit techniques, such as the following:

- Statistical sampling: limiting the number of test cases to a statistical (representative) sample of the universe being tested
- Restricting the value of input parameters or limiting the scope or field of inquiry
- Scheduled testing: breaking the audit or test into halves, quarters, or eighths and scheduling it over a period of months or even years, instead of doing the entire audit or test at one time

A word of caution: every test shortcut has its price in terms of potential risk. Management must weight the concerns about potential (not actual) risks, as against actual cost. This is particularly true if the potential risk is insured, that is, perceived to be someone else's problem should anything drastic occur.

There is much that one can learn from testing techniques used by other disciplines, such as safety. As an example, Herbert H. Jacobs lists the attributes of an effective measurement (test) system, as follows[1]:

- Administratively feasible
- Adaptable to the range of characteristics to be evaluated
- Constant
- Quantifiable
- Sensitive to change
- Valid in relation to what it is supposed to represent
- Capable of duplication with the same results from the same items measured
- Objective, efficient, and free from error

In our research, we have been especially concerned at the almost total absence of a generally accepted set of practices for testing security programs and systems. This being the case, the security professional can selectively borrow from related disciplines and adapt their principles or practices to help solve some security problems. One such field—auditing—sets forth the following guidelines, which can be borrowed and used with little or no modification:

- An audit is part of management's control.
- Management is the planning, organization, direction, and control of activities to achieve desired goals.
- It is necessary in a successful business process to set policy, establish procedure, assign responsibility, institute an accountability system, and measure performance.

[1]H. H. Jacobs, Toward More Effective Safety Measurement System. In *Measurement of Safety Performance*, edited by W. Tarrants (New York: Garland, 1980).

- Exceptionally good levels of security performance are achieved when risk control is perceived as an important and integral part of planning, organization, and direction.
- Risk control management systems must be integrated into the mainstream of all management functions.
- There is usually a noticeable difference between published policy and procedure and what actually occurs in most organizations.
- Seldom is an activity as effectively managed as those responsible for it say it is.

Auditing risk control (security) programs can serve as an appraisal of management's performance in relation to established company policy and procedure. The basic objective of the audit (test) is, however, the qualitative analysis of the existing security system to determine whether performance is effective and acceptable.

As stated by a senior executive of M & M Protection Consultants, "It has been our experience in conducting audits of the effectiveness of hazard (risk) control programs that there are usually two such programs in place at every location—the one management thinks it has and the one it really has!" He concluded that a high degree of failure is implicit if the hazard (risk) control program that management really has is a great deal less effective than the one management thinks it has.

DEVELOP A PLAN OF ACTION

It is no longer acceptable conduct for security practitioners to ask themselves, "Are we testing the right things, or are we testing things right?" What is necessary is to develop a plan of action for submission to management to make sure the job is accomplished. Some suggestions are as follows:

- Review the existing test procedures, if any.
- What kinds of tests are being conducted, by whom, when, and where?
- Are tests cost-effective and proven to reduce (eliminate) identifiable risks?
- Are records of past tests conducted being maintained for future use?
- Are there better, less expensive tests available that can be adapted for use at this facility?
- Are tests being conducted that can be eliminated as no longer functional or effective?
- Can you identify high-risk areas within the organization that are not being tested (audited)? Can you develop a suggested test program for management's review and approval?
- For complicated tests (audits and surveys), would it be in the best interest of the company to invite an outside expert or consultant in to conduct a review? To give a second opinion?
- Is it within your capability to develop and implement formal testing policy and guidelines, in those areas where you have operational responsibility?

Risk control specialists must seek out existing testing systems and promote development of new ones through which the effectiveness of security programs can be measured. Performance examination is a necessary element of the security professional's job description. One should always strive to do the best job possible, recognizing that under the best of circumstances, any measurement or test that one may develop, adopt, or adapt will have shortcomings. Any test, however, is better than none. The axiom: "test it, don't trust it," is a safe course of action. Ignoring the problem in hope that a failure, which may cause a serious incident, may never occur, is unacceptable.

19 ⊞

The Security Consultant

"There are four broad areas of competence that a consultant must achieve in order to be considered qualified. The four areas are: experience, education, professional credentials, personal and interpersonal skills...."
—Charles A. Sennewald, CPP, *Security Consulting*, 3rd edition, Butterworth-Heinemann, 2004

IN-HOUSE VERSUS OUTSIDE ADVICE

Many companies call on outside consultants to perform studies, make evaluations, and offer recommendations for implementing or improving their security programs. Some companies have benefited from the experience and knowledge that consultants can bring to bear on problems encountered during surveys. Other companies have not benefited. Disappointments are a result of a number of factors. For one thing, employees sometimes regard an outsider as an interloper, a stranger, one who has no real feeling for the company or its employees. Rank-and-file employees as well as supervisors and line managers may be resentful and secretive, thus preventing the "outsider" from obtaining a full understanding of problems as they presently exist within the company. No matter how experienced the consultant may be, his or her first task, and it is often a difficult and time-consuming one, is to learn the intricacies of the company, its ingrained processes, procedures, and methods of operation. This is often referred to as the *corporate culture*. Absent a full understanding of the corporate culture, the consultant's recommendations, usually seen first by line managers in the form of a written report, may produce a less than positive reaction. Some line managers may spend more time defending the status quo than implementing what may be valid recommendations for improving their operations. Employees generally know that outside consultants charge large fees for work that may well be done, at substantially less cost, using inside resources. Additionally, some so-called security consultants represent manpower or hardware firms and are salespeople in the truest sense of the word and not consultants at all. The title of consultant has been misused perhaps more in the security field than in any other profession. The result of using a salesperson who is not a qualified consultant is that the client often ends up paying for more "security" (manpower or hardware) than is actually needed.

One example that I encountered during a survey will suffice to make this point. A financial corporation dealing in wholesaling precious metals—gold, silver, and platinum—was found to be utilizing 16 closed-circuit television (CCTV) cameras in a 3,000 square foot office area. One fixed CCTV camera was mounted on the ceiling of an interior corridor located about 25 feet from the security console. This CCTV camera was targeted on the security console area, which, among other things, contained 16 CCTV display monitors. The security officer at the console was monitoring one CCTV display screen that presented him with a video image of himself at work! This can hardly be considered a cost-effective use of CCTV for protecting assets!

Another technique that I like to cite is the oversubscribed contract guard service. Recommending the elimination of just one guard post (coverage 24 hours per day, 7 days per week) can save a client about $80,000 per year. This is more than the usual cost of hiring a consultant. Thus, the client receives all the other benefits a security consultant brings to the table while also saving money on manpower costs.

Security consultants can and do provide valuable services to their clients, provided the client does a reasonably good job of selecting the right consultant in the first place. As a onetime professional security consultant, one who earned his living by plying this trade, I would caution prospective users of consulting services to use the same solid business judgment and standards in selecting a security consultant that one would in selecting any other type of consultant. In order to do that, perhaps a brief look into the historical development of the security consultant will be a worthwhile journey.

The field of protection consulting is relatively new, perhaps not more than 70 years old. Protection consulting had its origins in the insurance industry, principally with regard to property (fire) protection. The field then grew, as a natural extension, into accident prevention (casualty) and safety consulting. Last, but certainly not least, came security (crime prevention) consulting.

Security consulting probably got its start just before the United States entered World War II, with the development of the defense industry and its secret and top secret projects. Originally the emphasis was on perimeter protection, access control, and document classification as the principal means to protect defense secrets. The requirements for a security program of some sort were contractual in nature; that is to say, adequate protection was deemed to be necessary before the facility would be considered safe for secret or top secret defense projects. It was only when an obvious flaw or hole in the security was detected that an outside inspector came in and analyzed the situation. The inspector's job was to make recommendations to improve the security sufficiently for the facility to retain its clearance for secret or top secret production.

It is probably safe to say that most security consulting assignments then were based on problems that had already occurred. Little, if any, thought was given to prevention. It was during the 1970s that professional security consulting came into existence. It was also about this time that enlightened developers, owners, and managers began to recognize that to increase efficiency and reduce cost, security had to be built into facility design and not added onto a building project as an afterthought. Today, it is not uncommon for architects to seek the services of qualified protection consultants to ensure that their final designs take into account the security requirements for the buildings or project under consideration. As such, we now see security consultants specializing in the business of design engineering.

Working with architects and engineers on complex design and construction projects is not a task to be assigned to an apprentice security consultant. Clearly a combination of education and experience leading to professional maturity is needed here. It is said that a wise man knows his limitations. In the consulting field, mistakes can be costly. One's professional reputation can suffer if one takes on a project for which one is not fully qualified, and fails. Huge industrial complexes, such as nuclear power generating facilities or a large hospital complex, will probably require the services of a team of consultants, because of the multifaceted and varied disciplines required to survey such complicated environments. In the team approach, consultants are selected because of their expertise in the particular fields or specialty for which their talents will be utilized, recognizing that no one consultant can be expert in all fields of endeavor having to do with security or any other discipline.

Using the team approach to consulting assignments can reduce time and expenses for most large projects. Often it is the only way some large projects can be managed, because of the many specialty areas encountered in these environments. No one security consultant should be expected to be an expert in all phases of security management, procedure, hardware, and electronics—though most, by necessity, have a general idea of the proper application of the various security systems that may be used under specific conditions.

WHY USE OUTSIDE SECURITY CONSULTANTS?

I recall a telephone conversation with a security professional who asked to be referred to a text or written guide to help him design an electronic access control system for a building under development that was to house a financial institution's data processing center. At that time I knew of no such textbook (there have since been published a number of excellent texts on the subject of design security). I asked him, "Why don't you contract the job out to a security consultant with design engineering experience?" He stated, "I can't do that! My boss expects me to be able to handle every security problem that comes up, regardless of how complicated. I would be putting my professional reputation on the line if I ever admitted I didn't have the skills necessary to design an access control system." The simple answer to this situation is—nonsense! No professional from any discipline should be expected to be able to solve every problem that arises. This situation would be analogous to a general practitioner in the field of medicine calling a surgeon, stating that he or she had a patient who needed brain surgery for the removal of a tumor, then asking the surgeon to recommend a textbook so that general practitioner could read up on the subject before performing the operation himself or herself.

The above example is by no means uncommon. The question concerning when to use the services of an outside consultant does frequently arise. Some of the more common questions regarding the use of outside versus inside resources to do a security survey or consulting job are as follows.

Why Do I Need Outside Advice?

An independent consultant can furnish objective opinions without prejudice and without regard to internal pressures or politics. The consultant can, in effect, "let the chips fall where they may."

More often than not, a competent security director or manager knows what his problems are and has even defined the solutions. In these cases, the outside consultant can furnish a "second opinion" to reinforce the initial opinion, especially regarding cost-effective solutions to complicated problems.

When one seeks outside advice and assistance, one will surely seek help from a professional with a high degree of experience in dealing with the problems at hand. As mentioned earlier, the second opinion technique is common practice among other professions and disciplines. Yet in the security field, we find a great reluctance on the part of some professionals to admit to their obvious limitations.

Unlike manpower or hardware salesmen, the truly independent security professional has only one concern—the best interest of his client.

Manpower and hardware consultants (read salespeople) are generally limited in scope and are understandably biased toward their own products or services. Their first loyalty is to the company that employs them, and rightfully so. Nevertheless, security professionals have little reluctance in accepting proposals for service from contract security salespeople. The very same security professional will agonize over the prospect of recommending an outside security consultant to do a comprehensive security survey of his entire operation, including procedures, manpower, and hardware. So the next question often asked is the following:

How Can I Justify the Cost of a Consultant on a Limited Budget?

One must not lose sight of the fact that most security surveys are full-time propositions. Assuming that the in-house professionals are fully employed at their day-to-day occupations (and who in this business will admit that they are not?), where will they find the time to conduct a meaningful audit or survey?

Professional consultants usually have available to them library and research assistance unavailable to the average security practitioner. The library resources have been collected, catalogued, and indexed over a period of many years. Admittedly, with the advent of the Internet, this is less critical today than in the past.

Few security professionals, however, have developed the depth of knowledge necessary to do risk assessment in a multidisciplined environment. Most professionals tend to become specialists in certain fields—government, finance, utilities, hospitals, and retail, to name a few. It is not that most professionals are not capable of broadening their scope, it is just a fact of life that few of us do, preferring the "comfort" of our own field of expertise or practice.

An outside consultant can investigate the financial aspects of the necessary manpower and hardware solutions and then negotiate these cost factors with corporate management. Not every in-house security professional is schooled in the financial and negotiating techniques necessary to sell program changes. Most consultants, however, are.

Will an Outside Consultant Provide Assistance in Setting Up the Recommended Program?

This touches on a very common fear—that the consultant will make broad-brush recommendations and then walk off into the sunset, counting his or her excessive fee, leaving a difficult job for those who have to implement the consultant's recommendations. In actuality, consultants can continue to be employed to the extent that they and management feel is necessary to achieve the level of protection required to solve the problems identified during the survey. Risk assessment is at best a matter of opinion, with much uncertainty. The continued presence of the consultant with input at the implementation or installation stages can materially contribute to the final success of the project. And, physical presence is not always necessary. As we tell all our clients, "Night or day, we are only a telephone call away."

Most consultants do not provide contract services. Instead, they usually recommend several reliable firms in the immediate geographical vicinity that have reputations for providing quality service. The consultant then assists the client by drawing up minimum specifications and requirements, which the client furnishes to several firms, requesting that each submit a written bid. After the bids are returned to the client, the consultant can assist the client in reviewing the bids and selecting the service that meets the client's requirements at the best (not necessarily the lowest) cost. Once the service is accepted, the consultant can inspect, guide, provide administrative oversight, and critique the implementation or installation of the service.

This same procedure is applicable whether the product is security manpower, hardware, or electronics. But, as with all other phases of the survey, the consultant's key role is to function as the client's representative. Successful consultants function in the best interest of their clients at all times. This means scrupulously avoiding even the mere impression of a conflict of interest.

SECURITY PROPOSALS (WRITING AND COSTING)

A security survey can range from a simple telephone call, to a 1-day on-site review with verbal conclusions and recommendations, to a full field study. The last would encompass a comprehensive review of all risks, complete with a fully documented report detailing the entire security effort. Consulting assignments may also include plan development and review of blueprints and purchase specifications for access control and anti-intrusion alarm systems and other sophisticated security hardware and equipment.

To avoid misunderstanding the parameters of the task to be performed, both client and consultant should establish at the outset the specifications of the tasks to be performed. Probably the best way to accomplish this is with a written proposal.

Before a client asks for, or a consultant begins to prepare, a proposal, it is important that each have a basic understanding of the problems in need of being solved. This can be tricky. Often clients have only a limited idea of their problems and may not be able to articulate their needs. Some clients have not made a realistic appraisal of their problems and thus may not have realistic expectations regarding the solutions. The only way to ensure that both parties understand exactly what is to be accomplished is by outlining the issues in a written proposal.

Written proposals can take many forms, but five basic elements are common to most. They are the introduction, proposal, management, cost, and summary.

Introduction

The introduction identifies the client, geographical location, and problem in very broad terms. It also identifies the consultants and the firm that is submitting the proposal.

Proposal

The proposal must clearly state the need to be fulfilled, most often expressing it as a statement of work or scope. It sets forth in very specific terms both the problem and the proposed review or study that will be undertaken to gather the data necessary to solve the problem and meet the client's needs. It will also later serve as a general planning outline for the consultant doing the work.

Outlined below are some basic subject areas that may be considered in developing this part of the proposal. These areas are not all-inclusive and must be tailored or modified to fit the specifics of the task involved. They are presented here as examples only.

Security Objectives
There are usually four prime objectives that need to be developed during the evaluation of a facility: the risk assessment, vulnerability assessment, criticality assessment, and security function.

Losses
An in-depth assessment will be made of all incidents leading to losses at the site. Crime experience in the local area, investigation of existing shortages, and any incidences of fire, malicious damage, and vandalism will all be addressed in the assessment.

Security Organization
A review will be made of the security organization and structure as it pertains to vested authority, policy, assignment of responsibility, and cost-effectiveness.

Security Regulations and Procedures
A comprehensive review will be made of the security program in effect. This would include access control, personnel identification, package inspection, after-hours security procedures, liaison with police and other law enforcement agencies, and security indoctrination of employees.

Guard Force
A review of the present guard force or protective service will be made to cover organization, cost-effectiveness, training, report writing, and the effective utilization of manpower.

Personnel Security
This phase will include a review of background screening of employees, use of badges and passes, and termination procedures.

Physical Security Conditions
The survey team will evaluate the physical conditions, including all aspects of peripheral and interior security and security of objects that are protected or may need protection. This evaluation will consider the present facility, temporary conditions during construction, and proposed expansion plans.

Utilities
A security examination and evaluation will be conducted of critical utility and power points, for example, gas, telephone, computer, sewerage, water, and electricity. Storage practices and related security provisions will also be included.

Construction of Security Facilities

Detailed advisory information will be provided to the architect and engineer concerning the methods of construction and the installation of equipment that affects security. Examples are guardhouses, vaults, computer rooms, anti-intrusion devices, electronic card-access systems, and CCTV.

Security Hardware

A locksmith will evaluate the existing security hardware, such as physical deterrents, locks, key scheduling, associated hardware, and installations. Recommendations will cover repair and replacement of existing equipment and suggested material for new construction.

Alarm Systems

Evaluation will be made of the existing system and subsystems, to include expansion and improvements that may require substantial systems additions or complete replacement. State-of-the-art system conformity and performance will be considered. Interior and exterior intrusion detection systems, fire-detection and fire-suppression systems, and building evacuation plans will be part of this task.

Communications

A security evaluation will be made of the existing and proposed communications networks. These will include wired interior systems, telephone, cell phones, Global Positioning System (GPS) and computer systems, radio facilities, and satellite networks.

Surveillance

Security monitoring by CCTV and still and motion camera photography will be evaluated as applied or considered for future applications.

Security and Fire Safety Hardware

Security containers, security hardware, locks, and products employed for life safety, fire control, and fire extinguishments will be evaluated.

Procurement

Methodology for procurement, including sourcing, cost estimates, and scheduling, will be provided. Successful implementation of any security program hinges largely on a well-defined and executed procurement contract.

Management

The management section of the proposal will identify and fully describe the consulting organization, its experience, its personnel, and if necessary, a sampling of client companies that may be used as references. In any event, management, administration resources, logistics involved, and capabilities should be spelled out in some detail and should fully qualify the consultant and firm for the task at hand. Usually included in this part of the proposal are biographical sketches of the consultants who will actually be performing the survey.

Cost

Cost figures are the best-guess estimate of the consultant doing the job. They are only a yardstick and are subject to change if the scope of the inquiry changes when the job is underway. Nevertheless, the client is entitled to a reasonably accurate estimate of the cost and to prompt notification when the job is underway if the scope (and thus the cost) is going to change. Some clients specifically outline the task to be accomplished and send the outline out for several firms to bid on, and then accept the return proposal with the lowest figure. This technique, found most often in government entities, is called a *request for proposal* (RFP). It is also used by large multinational corporations with well-structured purchasing departments. The cost proposal will generally include the following factors:

- Direct labor (manpower) cost
- Travel and expenses
- Miscellaneous cost, if any
- Overhead rate (usually in percentage)
- General and administration (includes reports)
- Total estimated cost
- Profit
- Total proposal cost

The wise consultant will also include a 10 percent contingency fee, based on the total cost figure, to take care of such unforeseen problems as the following:

- Potential delays on site
- Meetings before, during, and after the on-site work commences
- Responses to follow-up inquiries after the final report is submitted
- Other unanticipated cost connected with the project

Summary

The summary is used to highlight the details of the proposal, as set forth in the previous four sections. It also contains the total cost of the proposed project, as obtained from section four. This section should identify the benefits the survey hopes to accomplish in terms that even the most recalcitrant, bottom line–oriented, bean-counting executive can understand. It must leave the reader with the positive feeling of having just read a proposal prepared in a timely, efficient, and professional manner. A late, poorly prepared, and disjointed proposal is a reflection of what the future holds regarding the primary task. Don't expect more or less from a consultant's proposal than you would expect to receive for the principal task.

A proposal pricing worksheet (Table 19.1) is included to assist both consultants and clients in developing cost figures for submission with proposals.

EVALUATION PROPOSALS AND REPORTS

Charles Hayden, PE, CPP, retired, formerly of the San Francisco office of Marsh & McLennan Protection Consultants, developed a list of minimum criteria to be applied in evaluating proposals and reports prepared by consultants. The following criteria were

Table 19.1   Proposal Pricing Worksheet

Name & Title of Contact: _____
PROSPECT NAME AND ADDRESS _____
   Street _____ Phone No. _____
   City & State _____ Zip Code _____

DESCRIBE SPECIFIC SERVICE TO BE PROVIDED:
_____
_____
_____
_____

BILLING INSTRUCTIONS: _____
   Describe– _____
_____
_____
_____

   Type Report Desired– _____
   Date Report Desired– _____

QUOTATION GOOD UNTIL _____

TIME	TRAVEL & EXPENSES (T&E)
No. Locations Involved: _____	Travel Hours: _____
On-Site Hours*: (Add 25-50% for Foreign) _____	Travel Costs: _____
Report Preparation Hours*: _____	Air _____ Cabs _____
Ratio Guide for Report Prep. to On-Site	Rail _____ Car Rental _____
Simple Reviews 1:1	Bus _____ Personal Car _____
Complex Reviews 2:1	Lodging: _____
*Above Hrs. Converted to Cost: _____	Food: _____
Support (typist, etc.) Cost: _____	Misc.: _____
TOTAL TIME COST: _____	Note: Normal domestic travel will range $100 to $130/day
	TOTAL T&E: _____

INCIDENTALS

Films _____ Printing _____   Slides _____ Binders _____
Projector _____ Pictures _____   Tapes _____ Other _____

TOTAL: _____

TIME COST _____   Estimated by: _____
T&E COST _____
INCIDENTALS COST _____   DATE: _____
10% CONTINGENCY _____
   GRAND TOTAL _____   APPROVED BY: _____

submitted to and adopted by the client. They are reproduced here with the approval and permission of Mr. Hayden.

The report (proposal) should fully satisfy the purpose for which the evaluation was made. The objective(s) of the evaluation should be identified and achieved. The scope of the evaluation must be consistent with the purpose and objective(s). The methodology used must be stated and must ensure that all significant information is collected, collated, and analyzed. The documented qualifications of the consultant must be adequate to perform the task.

Conclusions drawn in the report must include:

    a. Application of appropriate standards, acceptable practice, and/or experience.
    b. Credible estimates of comparative risk (probability/time) and potential damage/loss.

Recommendations for abatement of risk must be appropriate and effective in regard to:

    a. Priority.
    b. Cost.
    c. Estimated reduction of risk and potential damage.

The proposal should set forth a reporting procedure. Will the reports be periodic or final? When (date) can the client expect the report to be submitted, how, by whom, and in what form? Remember—keep the language and the format of reports simple.

Appendices

Security Survey Work Sheets

This is a basic guide that may be used to assist in performing physical security surveys in most industrial settings. Questions have been prepared for the purpose of reducing the possibility of neglecting to review certain areas of importance and to assist in the gathering of material for the survey report. Although the list is comprehensive, it is not all-inclusive. Individual adaptation will almost always be necessary to fit a specific environment or special circumstance.

Attached as Annex A and B are some specific questions that pertain to hospitals, universities, and colleges.

General Questions Before Starting a Survey

- Date of survey.
- Interview with (name of decision maker).
- Number of copies of the survey report desired by client, to be sent to:
- Obtain plot plan. Plot the production flow on plot plan and establish direction of north.
- Position and title of all people to be interviewed.
- Correct name and address of the facility.
- Type of business or manufacturer.
- Square footage of production or manufacturing space.
- Property other than main facility to be surveyed is located at:
- Property known as:
- Property consists of:
- What activity is conducted here?
- Is there other local property that will not be surveyed? Why?
- If plot plan is not complete, sketch remainder of property to be surveyed.

Number of Employees

- Administrative—total number all shifts.
- Skilled and unskilled—total number on each shift:
  1st shift

2d shift
3d shift
- Maintenance/cleanup crew
- Normal shift schedule and break times
- Salaried:
1st shift
2d shift
3d shift
- What days of the week is manufacturing in process?
- Are employees authorized to leave plant during breaks?
- Are hourly employees in a bargaining unit?
- Are company guards in a bargaining unit?

Cafeteria

- Where is the cafeteria located?
- What are the hours of operation?
- Is it company or concession operated?
- What is security of proceeds from sales?
- What is security of foodstuffs?
- What is method of supply of foodstuffs?
- How are garbage and trash removed?
- Where is location of vending machines?
- Where is change maker, if any?

Credit Union

- Where is the credit union located?
- How is money secured?
- How are records secured?
- How is office secured?
- What are the hours of operation?
- How much cash is kept on hand during day and overnight?

Custodial Service

- Staffed by outside contractor or company employees?
- What hours do janitors actually start and complete work?
- Do they have keys in their possession?
- How is trash removed by them?
- Who, if anyone, controls removal?
- Who controls their entrance and exit?
- Are they supervised by any company employee?

Company Store

- Where is the company store located?
- What are hours of operation?
- What method is used to control stock?
- How is stock supplied from plant?
- Number of clerks working in store?
- How is cash handled?
- When and who performs inventories?
- How are proceeds from sales secured?
- How is the store secured?

Petty Cash or Funds on Hand

- In what office are funds kept?
- What is the normal amount?
- How are these funds secured?
- What is process for the control and security of containers?
- Who has general knowledge of funds on hand?

Classified Operations

- Is government classified work performed?
- What is the degree of classification?
- How are classified documents secured?
- What is security during manufacture?
- What is classification of finished product?
- Are government inspectors on premises?
- Is company classified research and development (R&D) performed?
- Is company classified work sensitive to industry?
- What degree of security is it given?
- What degree of security does it require?
- Where are the locations of the various processing and storage areas?

Theft Experience

- Office machines or records.
- Locker room incidents.
- Pilferage from employees' autos.
- Pilferage from vending machines.
- Pilferage from money changer.
- Thefts of company-owned safety equipment.
- Theft of tools.
- Theft of raw material and finished product.

- Are thefts systematic or casual?
- Have any definite patterns been established?
- Are background investigations conducted before employment of personnel?
- What category of personnel is investigated?
- What is the extent of investigations?

The foregoing questions, answered properly, will assist you in developing the degree of control required for various areas. This information can be secured through interviews. You should also have a working knowledge of the organizational and operational plan of the facility. Before starting your detailed examination and study, you should take a guided (orientation) tour of the facility to acquaint yourself with the physical setting. Take notes during this tour, as necessary.

I. Physical Description of the Facility
- Is the facility subject to natural disaster phenomena?
- Describe in detail the above if applicable.
- What major vehicular and railroad arteries serve this facility?
- How many wood-frame buildings? Describe and identify them.
- How many load-bearing brick buildings? Describe and identify them.
- How many light or heavy steel-frame buildings? Describe and identify them.
- How many reinforced concrete buildings? Describe and identify them.
- Are all buildings within one perimeter? If not, describe.

II. Perimeter Security
- Describe type of fence, walls, buildings, and physical perimeter barriers.
- Is fencing of acceptable height, design, and construction?
- What is present condition of all fencing?
- Is material stored near fencing?
- Are poles or trees near fencing? If so, is height of fence increased?
- Are there any small buildings near fencing? If so, is the height of fence increased?
- Does undergrowth exist along the fencing?
- Is there an adequate clear zone on both sides along fencing?
- Can vehicles drive up to fencing?
- Are windows of buildings on the perimeter properly secured?
- Is wire mesh on windows adequate for its purpose?
- Are there any sidewalk elevators at this facility? If so, are they properly secured when not in operation?
- How are sidewalk elevators secured during operation?
- Do storm sewers or utility tunnels breach the barrier?
- Are these sewers or tunnels adequately secured?
- Is the perimeter barrier regularly maintained and inspected?
- How many gates and doors are there on the perimeter?
- Number used by personnel (visitors, employees)?
- Number used by vehicles?

- Number used by railroad?
- How is each gate controlled?
- Are all gates adequately secured and operating properly?
- Are railroad gates supervised by the guard force during operations?
- How are the railroad gates controlled?
- Do swinging gates close without leaving a gap?
- Are gates not used secured and sealed properly?
- What is security control of opened gates?
- Are chains and locks of adequate strength used to secure gates?
- Are alarm devices installed at the gates?
- Is CCTV used to observe gates or other part of the perimeter?
- How many doors from buildings open onto the perimeter?
- What type are they—personnel or vehicular?
- How are they secured when not in use?
- What is security control when in use?
- How many emergency doors breach the perimeter?
- How are emergency doors secured to prevent unauthorized use?
- Are there any unprotected areas on the perimeter?
- What portion of the perimeter do guards observe while making rounds?

III. Building Security
  A. Offices
   - Where are the various administrative offices located?
   - When are offices locked?
   - Who is responsible to check security at end of day?
   - How and where are company records stored?
   - How are they secured?
   - Are vaults equipped with temperature thermostats (rate-of-rise, Pyro-Larm)?
   - Are offices equipped with sprinklers? Fire extinguishers?
   - Are central station and local alarms installed to protect safes, cabinets, etc.?
   - Are file cabinets locked?
   - Are individual offices locked?
   - Does the company have a secure computer file server or communications room?
   - What type of fire protection does the facility have installed?
  B. Plant
   - When and how are exterior doors locked?
   - When and how are dock doors locked?
   - Are individual plant offices locked?
   - Are warehouses apart from production area secured?
   - Are certain critical and vulnerable areas protected by alarms? What type?
   - What are these areas? What do they contain?
  C. Tool Room
   - Is one or more established?

- Departmental or central tool room?
- What is the method of control and receipt?
- How is tool room secured?

D. Locker Rooms
- What is basis of locker issue to individual?
- What is type of locker—wall or elevated-basket type?
- How are individual lockers secured?
- Does company furnish keys/locks?
- Who or what department controls keys/locks?
- What control methods are used?
- How and when are keys and locks issued and returned?
- Are issued uniforms kept in lockers?
- Are unannounced locker inspections made?
- Who conducts inspections and how often?

E. Special Areas That May Require Additional Attention
If the facility houses the following types of activities, they may require special individual inspection. Recommendations can be based on any or all of the applicable portions of the checklist. After the initial inspection tour, you should design a checklist applicable to any special areas encountered.
- Research and development areas.
- Laboratories.
- Storage areas for valuable, critical, or sensitive items.
- Finished-product test areas.
- Finished-product display areas.
- Vehicle parking garages apart from the facility.
- Vacant or used lofts, attics, etc.
- Mezzanines or sub-basements.
- Aircraft hangars, maintenance shops, and crew quarters.

IV. Security of Shipping and Receiving Areas
- Locate all shipping docks, vehicle and railroad.
- What are the hours of operation of docks?
- What is the method of transportation?
- What is the method of inventory control at docks?
- What is the method of control of classified items?
- What is the security of classified or "hot" items?
- What supervision is exercised at the docks?
- Are loaded and unloaded trucks sealed?
- Who is responsible for sealing vehicles?
- What type of seals are being used?
- How are truck drivers controlled?
- Is there a designated waiting room for truck drivers?
- Is it separated from company employees?
- Are areas open to people other than dock employees?
- Do guards presently supervise these areas? Is this necessary?
- What is the method of accounting for material received?
- Is shipping done by parcel post, UPS, or FedEx?

- What is the control at point of packaging?
- Who controls stamps or stamp machines?
- Who transports packages to post office?
- What is the method of transport to post office?
- Where is pickup point at plant?
- What controls are exercised over the transport vehicle?
- Are inspections of operations made presently?
- Who conducts these inspections and how often?
- Does the facility have ship-loading wharves or docks?
- Are contract longshoremen used?
- How do longshoremen get to and from the docks?
- If they pass through the facility, how are they controlled?
- How are ships' company personnel controlled when given liberty?
- Are any specific routes through the facility designated for long-shoremen and ship personnel?
- If so, how are they marked and are they used?
- Are these personnel escorted?
- If they are not escorted, what measures are taken to supervise them?
- Is there any way in which these personnel could be kept from passing through the facility?

V. Area Security
- Can guards observe outside areas from their patrol routes?
- Do guards expose themselves to attack?
- Are patrols staggered so patterns are not established?
- What products are stored in outside areas?
- Is parking allowed inside the perimeter?
- If so, are controls established and enforced?
- Where do employees, visitors, and officials park?
- What security and control are provided?
- Are parking lots adequately secured?
- Is there a trash dump on the premises?
- How is it secured from the public?
- Is it manned by company employees?
- Is its approach directly from the manufacturing facility?
- Do roads within the perimeter present a traffic problem?
- Do rivers, canals, public thoroughfares, or railroads pass through the plant?
- Are loaded trucks left parked within the perimeter?
- If so, what protection is given them?
- Do the roads outside the facility present a traffic problem?
- What are these problems, and how can they be remedied?
- Is there any recreational activity within the perimeter, such as baseball?
- Are these areas fenced off from the remainder of the property?
- Could they logically be fenced off?

VI. Protective Lighting
- Is protective lighting adequate on perimeter?

- What type of lighting is it?
- Is lighting of open areas within perimeter adequate?
- Do shadowed areas exist?
- Are outside storage areas adequately lighted?
- Are inside areas adequately lighted?
- Is the guard protected or exposed by the lighting?
- Are gates adequately lighted?
- Do lights at gates illuminate the interior of vehicles?
- Are critical and vulnerable areas well illuminated?
- Is protective lighting operated manually or automatically?
- Do cones of light on perimeter overlap?
- Are perimeter lights wired in series?
- Is the lighting at shipping and receiving docks or piers adequate?
- Is lighting in the parking lots adequate?
- Is there an auxiliary power source available?
- Is the interior of buildings adequately lighted?
- Are top secret and secret activities adequately lighted?
- Are guards equipped with powerful flashlights?
- How much and what type of lighting is needed to provide adequate illumination? In what locations?
- Do security personnel report light outages?
- How soon are burned-out lights replaced?

VII. Key Control, Locking Devices, and Containers
- Is there a grandmaster, master, and submaster key system? Describe it.
- Are locks used throughout the facility of the same manufacture?
- Is there a record of issuance of locks?
- Is there a record of issuance and inspection of keys?
- How many grandmaster and master keys are there in existence?
- What is the security of grandmaster and master keys?
- What is the security of the key cabinet or box?
- Who is charged with handling key control? Is the system adequate? Describe the control system.
- What is the frequency of record and key inspections?
- Are keys made at the plant?
- Do keys have a special design?
- What is the type of lock used in the facility? Are locks adequate in construction?
- Would keys be difficult to duplicate?
- Are lock cores changed periodically at critical locations?
- Are any "sesame" padlocks used for classified material storage areas or containers?
- If a key cutting machine is used, is it properly secured?
- Are key blanks adequately secured?
- Are investigations made when master keys are lost?
- Are locks immediately replaced when keys are lost?

- Do locks have interchangeable cores?
- Are extra cores properly safeguarded?
- Are combination locks three-position type?
- Are safes located where guards can observe them on rounds?
- How many people possess combinations to safes and containers?
- How often are combinations changed?
- What type of security containers are used for the protection of money? Securities? High value metals? Company proprietary material? Government classified information?
- Are lazy-man combinations used?
- Are birth dates, marriage dates, etc., used as combinations?
- Are combinations recorded anywhere in the facility where they might be accessible to an intruder?
- Are combinations recorded and properly secured so that authorized individuals can get them in emergencies?
- Is the same or greater security afforded recorded combinations as that provided by the lock?
- Where government classified information is concerned, does each person in possession of a combination have the proper clearance and the "need to know"?
- Have all faces of the container locked with a combination lock been examined to see if combination is recorded?
- Are padlocks used on containers containing classified material chained to containers?

VIII. Control of Personnel and Vehicles
- Are passes or badges used? By whom?
- Type used? Describe in detail.
- Is color coding used?
- Are badges uniformly worn on outer clothing?
- Are special passes issued? To whom? When?
- Who is responsible for issue and receipt of passes and badges?
- Are badges and passes in stock adequately secured?
- How are outside contractors controlled?
- How are visitors controlled?
- How are vendors controlled?
- How many employee entrances are there?
- What type of physical control is there at each entrance and exit?
- Where are the time clocks located?
- Is it possible to consolidate clock locations to one or two main clock alleys?
- Is there any control at time clock locations?
- Are there special entrances for people other than employees?
- How are the special entrances controlled?
- Are fire stairwells used for operational purposes?
- Does the facility have elevators to access various floors?
- What control is exercised over their use?

- Are elevators used by operating employees?
- Do the elevators connect operational floors and office floors?
- Does this present a problem in personnel control?
- Are the elevators automatic or self-operated?
- If automatic, are floor directories posted in them?
- Do avenues within the buildings used for emergency egress present a problem of personnel control?
- Examine pedestrian flow from entrance, to locker room, and to work area.
- Can changes be made to shorten routes or improve control of personnel in transit?
- Are personnel using unauthorized entrances and exits?
- If government classified work is being performed, do controls in use comply with the Defense Contract Agency requirements for safeguarding classified information?
- Are groups authorized to visit and observe operations?
- How are these groups controlled?
- Are registers used to register visitors, vendors, and non-employees? Do they contain adequate information?
- Are these registers regularly inspected? By whom?
- Are employees issued uniforms?
- Are different colors used for different departments?
- What control is exercised over employees during lunch and coffee breaks?
- Do guards or watchmen ever accompany trash trucks or vending machine servicemen?
- Is parking authorized on premises within the perimeter?
- Are parking lots fenced off from the production areas?
- What method of control of personnel and vehicles is there in the parking lots?
- Is vehicle identification used?
- What type of vehicle stickers or identification is used?
- How are issue and receipt of stickers controlled?
- If executives park within the perimeter, are their autos exposed to employees?
- If nurses and doctors park within the perimeter, are their autos exposed to employees?
- Where do vendor servicemen park?
- Do vendor servicemen use plant vehicles to make the service tours? Are small vehicles available?
- How are outside-contractor vehicles controlled?
- What method is used to control shipping and receiving trucks?
- Are the parking facilities adequate at the docks?
- Does parking present a problem in vehicle or personnel control?
- What is the problem encountered?
- During what hour does switching of railroad cars occur?
- Is it possible for people to enter the premises during switching?

- Are there adequate directional signs to direct people to specific activities?
- Are the various buildings and activities adequately marked to preclude people from becoming lost?
- Are safety helmets required?
- Are safety shoes required?
- Are safety glasses required?
- Are safety gloves required?
- Are safety aprons required?
- Are full-time nurses or doctors available?
- Is there a vehicle available for emergency evacuation? What type is it?

IX. Safety for Personnel
- How far away is the nearest hospital in time and distance?
- Are any company employees or guards trained in first aid?
- Is a safety director appointed?
- Is there a safety program? What does it consist of?
- How often does the safety committee meet?
- Is a first aid or medical room available?
- How are medicine cabinets secured?
- Who controls these keys?
- How is the first aid room secured?
- Are any narcotics on hand?
- If so, has narcotics security been established?
- Are the required safety equipment items worn by employees? By visitors?
- What is the safety record of this facility?
- How does it compare with the national record?
- Are areas around machinery well policed?
- Does machinery have installed guards where needed? Are they used?
- Are mirrors used where needed to allow forklift operators to observe "blind" turns?
- Could or would mechanical devices used for forklift control improve safety?
- What type of device could be used? Pneumatic alarm system? Signal light?

X. Organization for Emergency
- Are doors adequate in number for speedy evacuation?
- Are they kept clear of obstructions and well marked?
- Are exit aisles clear of obstructions and well marked?
- Are emergency shutdown procedures developed, and is the evacuation plan in writing and periodically updated?
- Do employees understand the plans?
- Are emergency evacuation drills conducted?
- Do guards have specific emergency duties? Do they know these duties?
- Are local police available to assist in emergencies?

- Are any areas of the building in this facility designated as public disaster shelters?
- If so, are controls established to isolate the area from the rest of the facility?
- Do the emergency plans provide for a designated repair crew? Is the crew adequately equipped and trained?
- Are shelters available and marked for use of employees?
- If the plant is subject to natural disaster phenomena, what are they? Floods? Tornados? Hurricanes? Earthquakes?
- What emergency plans have been formulated to cope with potential hazards?
- When and what was the latest incident involving a natural disaster?
- Did it result in loss of life or loss or property?
- Attach a copy of the emergency procedures.

XI. Theft Control

- Are lunchbox inspections conducted?
- Is a package-pass control system being used? Describe it.
- Is a company-employed supervisor assigned to check the package-pass system regularly?
- Is a company official occasionally present during lunchbox inspections?
- Are package passes serially numbered or otherwise containing control numbers?
- Is security of package passes in stock adequate?
- Are comparison signatures available for inspection?
- Is the list of signatures kept up-to-date?
- What action is taken when employees are caught stealing?
- What controls are established on tools loaned to employees?
- What controls are established on laundry being removed?
- What is the method of removal of scrap and salvage?
- What controls are exercised over removal of useable scrap?
- Is control of this removal adequate?
- Are vending and service vehicle inspections being conducted?
- Do employees carry lunch boxes to their work areas?
- Are railroad cars inspected entering and leaving the plant?
- Are company-owned delivery or passenger vehicles authorized to park inside buildings of the facility?
- Does this parking constitute a possible theft problem?
- Do guards check outside the perimeter area for property thrown over fences?
- Do guards occasionally inspect trash pickup? Does anyone?

XII. Security Guard Forces

- What is present guard coverage—hours per day and total hours per week?
- Describe in detail guard organization and composition.
- Number and times of shifts each 24-hour period during weekdays and weekends?

- Number of stationary posts? When are they manned?
- Number of patrol routes? When and where are they, and when are patrols made?
- Are tours supervised? How?
- How many tour stations? Locate them on your plot plan (use different colors or shapes or symbols for different floors and routes).
- What is length of time of each tour?
- Is there additional coverage on Saturdays, Sundays, and holidays?
- Do the patrol routes furnish adequate protection for guards?
- Are the guards required to be deputized?
- Are armed guards required?
- How do guards communicate while on tour?
- Are written guard instructions available? If so, secure a copy.
- If no written instructions are available, generally describe duties of each shift and post.
- What equipment does the guard force have issued? Need?
- Do they require security clearances? What degree?
- Do they require special training?
- Is there a training program in force?
- What communications are available to the guard force to call outside the facility?
- Is the number of guards, posts, and tour routes adequate?
- Are mechanical or electronic devices used by the guard force?
- Do the guards know how to operate, reset, and monitor the devices properly?
- Do the guards know how to respond when alarms are activated?
- Are guard duties included in the emergency plans?
- Do guards know their duties? Emergency duties?
- Do guards make written reports of incidents?
- Are adequate records of incidents maintained?
- Are the guards familiar with the use of fire-fighting equipment?

Recommendations for changes must indicate each post, tour, and so forth, by number of hours for weekdays, weekends, and holidays, as well as a brief description of the guard's duties. List the total hourly guard service coverage at present and the total coverage after recommendations. If your recommendations increase guard service coverage, you must justify the hours and the cost.

SOME REFERENCE MATERIALS

- DOD Industrial Security Manual—Classified Information
- American Standard Practices for Protection Lighting
- General Electric Brochure, How to Select and Apply Floodlights
- Factory Mutual, Organizing Your Plant for Fire Safety
- Security Equipment Brochure Catalogue
- Alarm Installation Estimate Work Sheets

- NFPA—Quarterly Reports
- National Safety Council—Previous and Current Reports

ANNEX A: HOSPITAL SURVEYS

(Use applicable portions of the basic industrial security survey work sheet.)

A. Pharmacy
  - Where is the pharmacy located?
  - What are operating hours of pharmacy?
  - Is pharmacist registered and licensed?
  - Is license displayed in pharmacy?
  - How many pharmacists are employed?
  - How are narcotics received and recorded?
  - How are narcotics issued and recorded?
  - How are narcotics secured in pharmacy?
  - How are narcotics secured by nurses on wards?
  - How are narcotics secured in emergency room?
  - Are medicines issued by prescription only?
  - What type of inventory control is used for accounting?
  - What type of personnel control is used at the pharmacy?
  - Can entrance be easily gained, or is "Dutch door" used?
  - How are keys to pharmacy secured?
  - Are keys carried away from hospital by pharmacists?
  - Are "reach through" storage cupboards used?
  - Are "reach through" refrigerators used? If so, how are outside doors secured? Who possesses the keys?
B. Morgue
  - Where is the morgue located?
  - Does morgue remain locked when not in use?
  - Who is responsible for morgue security?
  - Who inventories items found on cadavers?
  - How are they inventoried and secured?
  - Who is authorized to release cadavers to undertakers?
  - Are local police escorted when they enter morgue?
C. Linen Department
  - Where is the linen department located?
  - What type of inventory control is used?
  - Is linen laundered on property?
  - Where is the laundry located?
  - How are various items marked for identification?
  - Is soiled linen accounted for upon receipt?
  - Is clean linen issued by receipt?
  - Are both the laundry and linen departments adequately secured?
D. Security of Receipts
  - Where are daily receipts paid and stored?
  - How are receipts secured?

- How much cash is normally accrued on one business day?
- How often and how is it deposited?
- Are containers furnished to secure patients' valuables?
- How are these valuables inventoried and secured?

E. Emergency Room
- Are security guards present?
- Do local police remain with patients they bring in?
- How are patients under the influence of alcohol or narcotics controlled?
- Are emergency medicines and narcotics properly secured?
- Who inventories valuables of patients arriving unconscious?
- How is this inventory done and when?
- Is a female nurse assigned or on call?
- Is the emergency entrance clear of obstructions?
- Is emergency vehicular approach kept clear?
- Are emergency phone numbers posted at or near the telephone?

F. Security Furnished to Nurses
- Where are the nurses' quarters located?
- Are nurses escorted to their quarters?
- Are nurses escorted to parking lots or to local transportation? If so, what time does this occur, and who escorts them?
- Are nurses' quarters included in guard tour system?

G. Security of Resident Doctors' Quarters
- Do doctors reside in the hospital?
- What is location of their quarters?
- How are keys to quarters issued?
- Are visitors allowed in quarters?
- Are doctors' quarters included in guard tour system?
- What is theft experience at quarters, if any?
- Are visiting doctors furnished residential quarters?
- What security is provided to residential quarters?
- Do doctors normally leave their medical bags unattended?
- What is theft experience at this location, if any?
- What security is exercised over doctors' parking area?
- Are signs displayed reminding them to lock their cars?

H. Dietary Department
- Where is the dietary department (kitchen) located?
- What type of inventory system is used?
- When are "dry" or canned goods received?
- When are fresh meats received?
- Who inventories foods and supplies received?
- How is food issued for preparation?
- What are hours of preparation?
- How is food for breakfast meal issued?
- What are the hours of operation of the cafeteria?
- How are personnel authorized to use cafeteria identified?
- What is percentage of turnover of dietary employees?

- Are these employees' backgrounds investigated?
- What system is used to issue food to bed patients?
- Do any floors have individual kitchens? If so, what system is used to issue food from stock?
- How are stock rooms secured?
- Who has possession of keys? Who controls keys?
- How is garbage removed?
- Is garbage ever inspected upon removal?
- How is combustible trash removed?
- Is this trash inspected upon removal?
- Are employees allowed to bring parcels, packages, or briefcases to work?

I. Identification and Control of Visitors

- What are the authorized visiting hours?
- How many and what entrances are used?
- Are visitors issued passes? Do they register?
- Who issues passes or registers visitors?
- Are passes required to be returned?
- Are passes color-coded by location within the hospital?
- How many visitors are authorized per patient at one time?
- Are visitors policed by the hospital staff?
- What system is used to prompt visitors to leave?
- What are the areas or locations where visitors are not authorized to go?

J. Emergency Evacuation Plan for Patients

- Have emergency evacuation plans been formulated?
- Are emergency evacuation drills conducted?
- Are staff members familiar with the plans?
- Are the procedures posted in strategic locations?
- Are emergency exits plainly marked?
- How are ambulatory patients evacuated?
- How are bed patients evacuated?
- Does the guard force participate in the evacuation of bed patients?
- Where are patients housed after they are evacuated?
- Are nurses, hourly employees, and staff members trained in emergency removal of bed patients?
- Are all hospital employees included in the emergency evacuation plan?
- Are local police and fire department personnel included in evacuation of operations?
- Does the emergency evacuation plan include specific routes for specific cases?
- Do the routes used for evacuation of bed patients conflict with the routes used by ambulatory patients?
- What system is used to sound the alert for fire?
- Is the method coded or disguised so as not to cause panic?

K. Parking Facilities
- Are the parking lots for doctors, staff, employees, and visitors separated? Posted?
- Is employee parking adjacent to the linen or dietary departments?
- If so, are fences erected so employees cannot gain entrance to their automobiles at will?
- Does the problem exist of visitors and/or doctors parking in fire and emergency lanes?
- Is emergency parking available for doctors?
- Does the administrator have a parking violation ticket system established? Describe it.
- Could visitors entering the hospital from the parking lot be canalized to pass a guard?
- Could this be accomplished for visitors entering from the street?
- Examine closely the traffic flow plan. Is it possible to organize the flow pattern and obtain a more efficient flow using one-way arteries or other devices?
- Is it possible to change the designation of the various parking areas (for instance, doctors' parking where visitors now park, and vice versa)?

ANNEX B: UNIVERSITY AND COLLEGE SURVEYS

A. Locking, Key Control, and Security Containers
   Note: All questions included in the basic industrial security survey work sheet will apply to universities and colleges, but the locking problems of such institutions are more complex and need additional scrutiny.
- Is there a system used for issuing keys to lockers to students at registration time?
- Is there a deposit charge made on the key?
- How much is the deposit?
- What type of lock is used on student lockers?
- How much difficulty is encountered if a lock must be changed when the student fails to return a key?
- Does the institution keep a record of keys issued to lockers?
- Are keys issued to students to any classrooms?
- Who keeps a record of such issue?
- Is there a deposit charge made on these keys?
- How much?
- Are locks changed between terms when such keys are not returned?
- Is any method provided to deny entrance to classrooms during hours when no one has authority to be in them?
- What protection is provided over examinations to be given?
- Is this material kept in combination or key-locked containers?

- If key locks are used, how securely is the room locked?
- Are windows to such rooms locked with key locks?
- Are windows to such rooms accessible from the ground or fire escapes?
- Is any special locking protection provided in areas where damage could be done to important experiments?
- What type of lock is used to protect areas where college funds are kept, bookstores, food storage areas, cafeteria offices, supply and equipment rooms, libraries, museum areas, valuable collections, dispensary narcotics, cabinets, and so forth?
- Are they adequate?
- How are doors to girls' dormitories locked and controlled?
- How are windows at first-floor level, basements, and those accessible by use of fire escapes protected?

B. Protective Lighting
- Are open areas of the campus sufficiently lighted to discourage illegal or criminal acts against pedestrians?
- Is the campus equipped with emergency call stations placed at strategic locations and in open areas?
- Are there any areas that are covered with high growing shrubs or trees where the light is not sufficient?
- Are the outsides of buildings holding valuable or critical activities or materials lighted?
- Are interiors of hallways and entrances lighted when buildings are open at night?
- Are areas surrounding the dormitories and living quarters well lighted?
- Are lighting fixtures used for this purpose placed in a position to shine out from the building into the eyes of a person approaching?
- Are halls, building entrances, and dormitories well lighted?
- Are campus parking lots lighted sufficiently to discourage tampering with parked cars or other illegal activities?
- Are areas where materials of high value are stored well lighted? Safes, libraries, bookstores, food storage areas, and the like?

APPENDIX B

Danger Signs of Fraud, Embezzlement, and Employee Theft[1]

I. Situational Pressures
    A. Individuals Against the Company
        1. Do any of the key employees have unusually high personal debts or financial losses (i.e., high enough that they probably could not meet them with their own level of income)?
        2. Do any of the key employees appear to be receiving incomes that are inadequate to cover normal personal and family expenses?
        3. Do any of the key employees appear to be living beyond their means?
        4. Are any of the key employees involved in extensive stock market or other speculation (i.e., extensive enough so that a downturn would cause them severe financial difficulty)?
        5. Are any of the key employees involved in excessive or habitual gambling?
        6. Do any of the key employees have unusually high expenses resulting from personal involvement with other people (e.g., maintaining separate apartments or households)?
        7. Do any of the key employees feel undue family, community, or social expectations or pressures?
        8. Do any of the key employees use alcohol or drugs excessively?
        9. Do any of the key employees strongly believe that they are being treated unfairly (e.g., underpaid, poor job assignments)?

---

[1]From *How to Detect and Prevent Business Fraud* by W. Steve Albrecht, PhD, CPA, Marshall B. Romney, PhD, CPA, David J. Cherrington, DBA, I. Reed Payne, PhD, Allan J. Roe, PhD. © 1982 by Prentice-Hall, Inc. Published by Prentice-Hall, Inc., Englewood Cliffs, N.J., 07632.

10. Do any of the key employees appear to resent their superiors?

11. Are any of the key employees unduly frustrated with their jobs?

12. Is there an undue amount of peer pressure to achieve in this company, so much so that success is more important than ethics?

13. Do any of the key employees appear to exhibit extreme greed or an overwhelming desire for self-enrichment or personal gain?

B. Individuals on Behalf of the Company

1. Has the company recently experienced severe losses from any major investments or ventures?

2. Is the company attempting to operate with insufficient working capital?

3. Does the company have unusually high debts, so high that either interest payments or balances due impose a threat to the stability of the company?

4. Have tight credit or high interest rates reduced the company's ability to acquire credit? Is there undue pressure to finance expansion through current earnings rather than through debt or equity?

5. Is the company caught in a profit squeeze (i.e., are costs and expenses rising higher and faster than sales and revenues)?

6. Do existing loan agreements provide little available tolerance on debt restrictions?

7. Has the company's quality of earnings been progressively deteriorating (e.g., adopting straight-line depreciation to replace an accelerated depreciation without good reason, or reporting good profits but experiencing cash shortage)?

8. Is the company experiencing an urgent need to report favorable earnings (e.g., to support a high stock price or to meet forecasted earnings)?

9. Does company management believe there is a need to gloss over a "temporarily bad situation" in order to maintain management position and prestige?

10. Does the company have a significant amount of unmarketable collateral?

11. Does the company depend heavily on only one or two products, customers, or transactions?

12. Does the company have an excess of idle productive capacity?

13. Does the company suffer from severe obsolescence (i.e., is a significant amount of inventory or physical facilities obsolete)?

14. Does the company have an unusually long business cycle, long enough so that profits or cash flows are threatened?

15. Does the company have any revocable or possibly imperiled licenses that are necessary for the firm's existence or continued operation?

16. Has the company expanded rapidly through new business or product lines? If so, has expansion been orderly or has it been done in an attempt to salvage profitability?

17. Are there currently, or have there recently been, unfavorable economic conditions within this company's industry, or is the company's performance running counter to industry trends?

18. Is the company experiencing undue difficulty in collecting receivables (i.e., is the receivable turnover slowing down)?

19. Does the company face unusually heavy competition, heavy enough that its existence appears threatened?

20. Is the company experiencing a significant reduction in sales backlog indicating a future decline in sales?

21. Is the company being pressured to either sell out or merge with another company?

22. Is the company experiencing sizable inventory increases without comparable sales increases?

23. Has the company recently experienced any significant adverse tax adjustments or changes?

24. Is the company experiencing significant litigation, especially between stockholders and management?

25. Has the company recently been suspended or delisted from a stock exchange?

II. Opportunities

A. Individuals Against the Company

1. Do any of the key employees have close associations with suppliers or key individuals who might have motives inconsistent with the company's welfare?

2. Does the company fail to inform employees about rules of personal conduct and the discipline of fraud perpetrators?

3. Is the company experiencing a rapid turnover of key employees, through their quitting or being fired?

4. Have any of the key employees recently failed to take annual vacations of more than 1 or 2 days, or has the company failed periodically to rotate or transfer key personnel?

5. Does the company have inadequate personnel screening policies when hiring new employees to fill positions of trust (e.g., check on secondary references, etc.)?

6. Does the company lack explicit and uniform personnel policies?

7. Does the company fail to maintain accurate personnel records of dishonest acts or disciplinary actions?

8. Does the company fail to require executive disclosures and examinations (e.g., personal investments or incomes)?

9. Does the company appear to have dishonest or unethical management?
10. Is the company dominated by only one or two individuals?
11. Does the company appear to operate continually on a crisis basis?
12. Does the company fail to pay attention to details (e.g., are accurate accounting records unimportant)?
13. Does the company place too much trust in key employees and overlook traditional controls?
14. Is there a lack of good interpersonal relationships among the key executives in the company?
15. Does the company have unrealistic productivity measurements or expectations?
16. Does the company have poor compensation practices? Is pay commensurate with the level of responsibility?
17. Does the company lack a good system of internal security (e.g., locks, safes, fences, gates, and guards)?
18. Does the company have adequate training programs?
19. Does the company have an inadequate internal control system, or does it fail to enforce the existing controls?

B. Individuals on Behalf of the Company
   1. Has the company recently had any significant related-party transactions?
   2. Does the company retain different auditing firms for major subsidiaries, or does it change auditors often?
   3. Is the company reluctant to provide the auditors with data needed to complete the audit examination?
   4. Does the company retain several different legal counsels, or does it change legal counsels often?
   5. Does the company use several different banks, none of which can see the company's entire financial picture?
   6. Does the company seem to have continuous problems with regulatory agencies?
   7. Does the company possess an unduly complex business structure, so complex that many facets lack purpose or meaning?
   8. Does the company seem to need but lack an effective internal auditing staff?
   9. Is the company highly computerized? If so, are there insufficient controls over hardware, software, computer personnel, etc.?
   10. Does the company have an inadequate internal control system, or does it fail to enforce the existing internal controls?
   11. Is the company in a "hot" or high-risk industry (i.e., an industry that has experienced a large number of business failures or frauds)?
   12. Does the company have a number of large year-end or unusual transactions?

13. Does the company have unduly liberal accounting practices?
14. Does the company have poor accounting records?
15. Does the accounting department of the company appear to be inadequately staffed?
16. Does the company fail to disclose questionable or unusual accounting practices?

C. Personal Characteristics

1. Do any of the key employees appear to have low moral character?
2. When confronted with difficulty, do any of the key employees appear consistently to rationalize contradictory behavior?
3. Do any of the key employees appear to lack a strong personal code of honesty?
4. Do any of the key employees appear to be "wheeler-dealer" individuals who enjoy feelings of power, influence, social status, and excitement associated with financial transactions involving large sums of money?
5. Do any of the key employees appear to be unstable (e.g., frequent job changes, frequent changes of residence, mental problems)?
6. Do any of the key employees appear to be intrigued by the personal challenge of subverting a system of controls (i.e., do they appear to have a desire to beat the system)?
7. Do any of the key employees have criminal or questionable backgrounds?
8. Do any of the key employees have poor credit ratings?
9. Do any of the key employees have poor past work records or references?

APPENDIX C

Professional Practices for Business Continuity Planners

The following material is provided by the Disaster Recovery Institute International (DRII) and represents their Common Body of Knowledge for the business continuity planning field. For additional information, contact:

> DRI International
> 201 Park Washington Court
> Falls Church, VA 22046-4513
> Phone: 703-538-1792
> http://www.drii.org

INTRODUCTION

Professions are characterized by a body of knowledge shared by members of the profession and used in their work. Specific skills, tasks, or activities for the profession emerge and evolve from a set of subject areas of a common body of knowledge that characterize the profession. In the business continuity planning profession, this common body of knowledge is the Professional Practices for the Business Continuity Planner. This body of knowledge is accepted by both DRII and by the Business Continuity Institute (BCI) based in the United Kingdom.

The existence of such a body of knowledge is necessary but is not sufficient evidence of the existence of the profession. General acceptance requires proper application and periodic updates to the body of knowledge for success in the profession. Both DRII and BCI are committed to joint maintenance and acceptance of the Professional Practices.

This document defines the boundaries of the business continuity planning profession and the base of knowledge that qualifies one for DRII certification as an Associate Business Continuity Planner (ABCP), Certified Business Continuity Professional (CBCP), or Master Business Continuity Professional (MBCP). Likewise, the BCI uses the Professional Practices as the basis for their examination procedures for Membership of the Business Continuity Institute (MBCI) and Fellowship of the Business Continuity Institute (FBCI).

For DRII certification purposes, practitioners must demonstrate continuing involvement and experience in business continuity planning, in addition to successful completion of a written examination based on the Professional Practices. Demonstrated experience must relate to the content of the common body of knowledge.

> Joint adoption of this body of knowledge by both DRII and BCI, effective August 28, 2003, recognizes the term *business continuity management* to define holistic management processes that identify potential impacts that threaten an organization and provide a framework for building resilience with the capability for an effective response that safeguards the interests of its key stakeholders, reputation, and value-creating activities.

The primary objective of business continuity management is to allow business operations to continue under adverse conditions, by the introduction of appropriate resilience strategies, recovery objectives, and business continuity and crisis management plans in collaboration with, or as a key component of, an integrated risk management initiative.

The 10 sections of these standards are not presented in any particular order of importance or sequence because it may be necessary to undertake or implement sections in parallel during the development of a business continuity management program.

Each subject area in this document provides the following:

- A description of the subject area
- The role of the professional
- An outline of the knowledge that the professional should demonstrate within each subject area

Illustrative examples and references are also provided where appropriate.

SUBJECT AREA OVERVIEW

1. Project Initiation and Management

Establish the need for a Business Continuity Management (BCM) Process or Function, including resilience strategies, recovery objectives, and business continuity and crisis management plans and including obtaining management support and organizing and managing the formulation of the function or process either in collaboration with, or as a key component of, an integrated risk management initiative.

2. Risk Evaluation and Control

Determine the events and external surroundings that can adversely affect the organization and its resources (facilities, technologies, etc.) with disruption as well as disaster, the damage such events can cause, and the controls needed to prevent or minimize the effects of potential loss. Provide cost/benefit analysis to justify investment in controls to mitigate risks.

3. Business Impact Analysis

Identify the impacts resulting from disruptions and disaster scenarios that can affect the organization and techniques that can be used to quantify and qualify such impacts. Identify time-critical functions, their recovery priorities, and interdependencies so that recovery time objectives can be set.

4. Developing Business Continuity Management Strategies

Determine and guide the selection of possible business operating strategies for continuation of business within the recovery point objective and recovery time objective, while maintaining the organization's critical functions.

5. Emergency Response and Operations

Develop and implement procedures for response and stabilizing the situation following an incident or event, including establishing and managing an Emergency Operations Center to be used as a command center during the emergency.

6. Developing and Implementing Business Continuity and Crisis Management Plans

Design, develop, and implement Business Continuity and Crisis Management Plans that provide continuity within the recovery time and recovery point objectives.

7. Awareness and Training Programs

Prepare a program to create and maintain corporate awareness and enhance the skills required to develop and implement the Business Continuity Management Program or process and its supporting activities.

8. Maintaining and Exercising Business Continuity Plans

Preplan and coordinate plan exercises and evaluate and document plan exercise results. Develop processes to maintain the currency of continuity capabilities and the plan document in accordance with the organization's strategic direction. Verify that the Plan will prove effective by comparison with a suitable standard, and report results in a clear and concise manner.

9. Crisis Communications

Develop, coordinate, evaluate, and exercise plans to communicate with internal stakeholders (employees, corporate management, etc.), external stakeholders (customers, shareholders, vendors, suppliers, etc.), and the media (print, radio, television, Internet, etc.).

10. Coordination with External Agencies

Establish applicable procedures and policies for coordinating continuity and restoration activities with external agencies (local, state, national, emergency responders, defense, etc.) while ensuring compliance with applicable statutes or regulations.

SUBJECT AREA 1: PROJECT INITIATION AND MANAGEMENT

Establish the need for Business Continuity Planning (BCP) within a BCM Process, including resilience strategies, recovery objectives, business continuity and crisis management plans, and including obtaining management support and organizing and managing the project to initiate the process to completion within agreed upon time and budget limits.

    A. The Professional's Role Is to:
        1. Lead Sponsors in Defining Objectives, Policies, and Critical Success Factors
            a. Scope and objectives
            b. Legal and requirements reasons
            c. Case histories and industry best practices
        2. Coordinate and Organize/Manage the BCP Project and Overall BCP Process using a steering committee and project task force
        3. Oversee the BCP Process Through Effective Control Methods and Change Management
        4. Present (Sell) the Process to Management and Staff
        5. Develop Project Plan and Budget to initiate the process
        6. Define and Recommend Process Structure and Management
        7. Manage the Project to Develop and Implement the BCP Process
    B. The Professional Should Demonstrate a Working Knowledge in the Following Areas:
        1. Establish the Need for Business Continuity
            a. Reference relevant legal/regulatory/statutory/contractual requirements and restrictions
            b. Reference relevant regulations of industry trade bodies or associations, where appropriate
            c. Reference current recommendations of relevant authorities
            d. Relate legislation, regulations, and recommendations to organizational policy
            e. Identify any conflicts between organizational policies and relevant external requirements
            f. Identify any audit records
            g. Propose methods, which may include a BCP or crisis management plan, to resolve any conflicts between organizational policies and relevant external requirements

    h. Identify business practices (e.g., just-in-time inventory) that may adversely impact the organization's ability to recover following a disaster event

2. Communicate the Need for a BCP
    a. Develop awareness by means of formal reports and presentations
    b. State the benefits of the BCP and relate the benefits to organizational mission, objectives, and operations
    c. Gain organizational commitment to the BCP process
    d. Develop a mission statement/charter for the BCP process

3. Involve Executive Management in the BCP Process
    a. Explain executive management's role in the BCP process
    b. Explain and communicate management's accountability and liability for the BCP process

4. Establish a Planning/Steering Committee: Roles and Responsibilities, Types of Organization, Control and Development, and Membership
    a. Select appropriate personnel
    b. Define their roles and responsibilities
    c. Develop a suitable set of objectives for the BCP process

5. Develop Budget Requirements
    a. Clearly define resource requirements
    b. Obtain estimates of financial requirements
    c. Verify the validity of resource requirements
    d. Validate the estimates of financial requirements
    e. Negotiate resource and financial requirements with management
    f. Obtain executive commitment for financial requirements

6. Identify Planning Team(s) and Responsibilities
    a. Emergency management/incident response/crisis management team
    b. Business continuity planning teams (multilocation, multidivision, etc.)
    c. Recovery/response and restoration teams

7. Develop and Coordinate Project Action Plans to Develop and Implement the BCP Process
    a. Develop an overall project plan with realistic time estimates and schedule

8. Develop the Ongoing Management and Documentation Requirements for the BCP Process

9. Report to Senior Management and Obtain Senior Management Approval/Commitment
    a. Set up a schedule to report the progress of the BCP process to senior managers
    b. Develop regular status reports for senior management that contain concise, pertinent, accurate, and timely information on key parameters of interest or information of which senior management should be made aware

SUBJECT AREA 2: RISK EVALUATION AND CONTROL

Determine the events and external surroundings that can adversely affect the organization and its facilities with disruption as well as disaster, the damage such events can cause, and the controls needed to prevent or minimize the effects of potential loss. Provide cost/benefit analysis to justify investment in controls to mitigate risks.

A. The Professional's Role Is to:
  1. Identify Potential Risks to the Organization
     a. Probability
     b. Consequences/Impact
  2. Understand the Function of Risk Reduction/Mitigation within the Organization
  3. Identify Outside Expertise Required
  4. Identify Exposures
  5. Identify Risk Reduction/Mitigation Alternatives
  6. Confirm with Management to Determine Acceptable Risk Levels
  7. Document and Present Findings
B. The Professional Should Demonstrate a Working Knowledge in the Following Areas:
  1. Understand Loss Potentials
     a. Identify exposures from both internal and external sources. These should include, but not be limited to, the following:
        (1) Natural, manmade, technological, or political disasters
        (2) Accidental versus intentional
        (3) Internal versus external
        (4) Controllable risks versus those beyond the organization's control
        (5) Events with prior warnings versus those with no prior warnings
     b. Determine the probability of events
        (1) Information sources
        (2) Credibility
     c. Create methods of information gathering
     d. Develop a suitable method to evaluate probability versus severity
     e. Establish ongoing support of evaluation process
     f. Identify relevant regulatory and/or legislative issues
     g. Establish cost/benefit analysis to be associated with the identified loss potential
  2. Determine the Organization's Exposures to Loss Potentials
     a. Identify primary exposures the organization may face, and secondary/collateral events that could materialize because of such exposures (e.g., hurricane threat could result in several events including high winds, flood, fire, building and roof collapse, etc.)
     b. Select exposures most likely to occur and with greatest impact

3. Identify Controls and Safeguards to Prevent and/or Mitigate the Effect of the Loss Potential
   Considerations: The actions taken to reduce the probability of occurrence of incidents that would impair the ability to conduct business.
   a. Physical protection
      (1) Understand the need to restrict access to buildings, rooms, and other enclosures where circumstances demand a "three-dimensional" consideration
      (2) Understand the need for barriers and strengthened structures to determine wilful and accidental and/or unauthorized entry
      (3) Location: physical construction, geographic location, corporate neighbors, facilities infrastructure, community infrastructure
   b. Physical presence
      (1) Understand the need for the use of specialist personnel to conduct checks at key entry points
      (2) Understand the need for manned and/or recorded surveillance equipment to control access points and areas of exclusion, including detection, notification, suppression
      (3) Understand security and access controls, tenant insurance, leasehold agreements
   c. Logical protection
      (1) Understand the need for system-provided protection of data stored, in process, or in translation; information backup and protection
      (2) Understand detection, notification, suppression
      (3) Understand information security: hardware, software, data, network
   d. Location of assets
      (1) Understand the inherent protection afforded key assets by virtue of their location relative to sources of risk
      (2) Personnel procedures
      (3) Preventive maintenance and service as required
      (4) Utilities: duplication of utilities, built-in redundancies (telecommunications, power, water, etc.)
      (5) Interface with outside agencies (vendors, suppliers, outsourcers, etc.)
4. Evaluate, Select, and Use Appropriate Risk Analysis Methodologies and Tools
   a. Identify alternative risk analysis methodologies and tools
      (1) Qualitative and quantitative methodologies
      (2) Advantages and disadvantages

   (3) Reliability/confidence factor

   (4) Basis of mathematical formulas used

  b. Select appropriate methodology and tool(s) for company-wide implementation

5. Identify and Implement Information-Gathering Activities

  a. Develop a strategy consistent with business issues and organizational policy

  b. Develop a strategy that can be managed across business divisions and organizational locations

  c. Create organization-wide methods of information collection and distribution

   (1) Forms and questionnaires

   (2) Interviews

   (3) Meetings

   (4) Documentation review

   (5) Analysis

6. Evaluate the Effectiveness of Controls and Safeguards

  a. Develop communications flow with other internal departments/divisions and external service providers

  b. Establish business continuity service-level agreements for both supplier and customer organizations and groups within and external to the organization

  c. Develop preventive and preplanning options

   (1) Cost/benefit

   (2) Implementation priorities, procedures, and control

   (3) Testing

   (4) Audit functions and responsibilities

  d. Understand options for risk management and selection of appropriate or cost-effective response, i.e., risk avoidance, transfer, or acceptance of risk

7. Risk Evaluation and Control

  a. Establish disaster scenarios based on risks to which the organization is exposed. The disaster scenarios should be based on these criteria: severe in magnitude, occurring at the worst possible time, resulting in severe impairment to the organization's ability to conduct business.

  b. Evaluate risks and classify them according to relevant criteria, including risks under the organization's control, risks beyond the organization's control, exposures with prior warnings (such as tornados and hurricanes), and exposures with no prior warnings (such as earthquakes).

  c. Evaluate impact of risks and exposures on those factors essential for conducting business operations: availability of personnel, availability of information technology, availability of communications technology, status of infrastructure (including transportation), etc.

        d. Evaluate controls and recommend changes, if necessary, to reduce impact due to risks and exposures
           (1) Controls to inhibit impact exposures: preventive controls (such as passwords, smoke detectors, and firewalls)
           (2) Controls to compensate for impact of exposures: reactive controls (such as hot sites)

    8. Security
        a. Identify the organization's possible security exposures, including the following specific categories of security risks
           (1) Physical security of all premises
           (2) Information security—computer room and media storage area security
           (3) Communications security—voice and data communications security
           (4) Network security—intranet security, Internet security
           (5) Personnel security
        b. Advise on feasible, cost-effective security measures required to prevent/reduce security-related risks and exposures

    9. Vital Records Management
        a. Identify vital record needs in the organization, including paper and electronic records
        b. Evaluate existing backup and restoration procedures for vital records
        c. Advise on and implement feasible, cost-effective backup and restoration procedures for all forms of the organization's vital records

SUBJECT AREA 3: BUSINESS IMPACT ANALYSIS

Identify the impacts resulting from disruptions and disaster scenarios that can affect the organization and techniques that can be used to quantify and qualify such impacts. Establish critical functions, their recovery priorities, and interdependencies so that recovery time objective(s) and recovery point objective(s) can be set.

    A. The Professional's Role Is to:
        1. Identify Knowledgeable Functional Area Representatives for the Business Impact Analysis (BIA) process
        2. Identify Organization Functions Including Information and Resources (people, technology, facilities.)
        3. Identify and Define Criticality Criteria
        4. Obtain Management Approval for Criteria Defined
        5. Coordinate Analysis
        6. Identify Interdependencies (internal and external to the organization)
        7. Define Recovery Objectives and Time Frames

8. Define Report Format
9. Prepare and Present Final BIA to Management
B. The Professional Should Demonstrate a Working Knowledge in the Following Areas:
   1. Establish the Project
      a. Identify and obtain a project sponsor for the BIA activity
      b. Define objectives and scope for the BIA project
      c. Choose an appropriate BIA project planning methodology/tool
      d. Identify and inform participants of the BIA project and its purpose
      e. Identify training requirements, establish a training schedule, and undertake training as appropriate
      f. Obtain agreement on final project time schedule and initiate the BIA project
   2. Assess Effects of Disruptions, Loss Exposure, and Business Impact
      a. Effects of disruptions
         (1) Loss of assets: key personnel, physical assets, information assets, intangible assets
         (2) Disruption to the continuity of service and operations
         (3) Violation of law/regulation
         (4) Public perception
      b. Impact of disruptions on business
         (1) Financial
         (2) Customers and suppliers
         (3) Public relations/credibility
         (4) Legal
         (5) Regulatory requirements/considerations
         (6) Environmental
         (7) Operational
         (8) Personnel
         (9) Other resources
      c. Determine loss exposure
         (1) Quantitative
            (a) Property loss
            (b) Revenue loss
            (c) Fines
            (d) Cash flow
            (e) Accounts receivable
            (f) Accounts payable
            (g) Legal liability
            (h) Human resources
            (i) Additional expenses/increased cost of working
         (2) Qualitative
            (a) Human resources
            (b) Morale
            (c) Confidence

(d) Legal

(e) Social and corporate image

(f) Financial community credibility

3. Business Impact Analysis (BIA)—A Suggested Methodology: Understand Assessment Techniques: Quantitative and Qualitative Methods

   a. BIA data collection methodologies

      (1) Finalize an appropriate data collection method (e.g., questionnaires, interviews, workshop, or an agreed combination)

         (a) Data collection via questionnaires

            i. Understand the need for appropriate design and distribution of questionnaires, including explanation of purpose, to participating departmental managers and staff

            ii. Understand the role of and manage project kick-off meetings to distribute and explain the questionnaire

            iii. Understand the role of and support respondents during completion of questionnaires

            iv. Review completed questionnaires and identify those requiring follow-up interviews

            v. Conduct follow-up discussions when clarification and/or additional data are required

         (b) Data collection via interviews only

            i. Understand the need for consistency, with the structure of each interview predefined and following a common format

            ii. Ensure the base data to be collected at each interview are predefined

            iii. Understand the need for initial interview to be reviewed and verified by the interviewee

            iv. Schedule follow-up interviews, if initial analysis shows a need to clarify and/or add to the data already provided

         (c) Data collection via workshop

            i. Understand the need for and establish a clear agenda and set of objectives

            ii. Identify the appropriate level of participating management and obtain agreement

            iii. Choose appropriate venue, evaluating location, facilities, and staff availability

            iv. Act as facilitator and leader during discussions

            v. Ensure workshop objectives are met

            vi. Ensure all issues outstanding at the end of the workshop are identified and responsibility for their resolution agreed on

      (2)   Recommend and obtain agreement as to how potential financial and nonfinancial impact can be quantified and evaluated

      (3)   Identify and obtain agreement on requirements for nonquantifiable impact information and gain agreement

      (4)   Develop questionnaire (if used) and completion instructions

      (5)   Determine data analysis methods (manual or computer)

   b. BIA report

      (1)   Prepare draft BIA report containing initial impact findings and issues

      (2)   Issue draft report to participating managers and request feedback

      (3)   Review manager feedback and, when appropriate, revise findings accordingly or add to outstanding issues

      (4)   Schedule a workshop or meeting with participating manager(s) to discuss initial findings, when necessary

      (5)   Ensure that original findings are updated to reflect changes arising from these meetings

      (6)   Prepare final BIA report according to organization

      (7)   Prepare and undertake formal presentation of BIA findings to peers and executive bodies
    Note: No standards exist for the format or distribution of BIA reports, so these reports will vary between organizations.

4. Define Criticality of Business Functions and Records, and Prioritize

   a. Establish definition of criticality, and negotiate with management either single or multiple levels of criticality

   b. Identify and prioritize critical functions

      (1)   Business functions

      (2)   Support functions

   c. Identify and prioritize vital records to support business continuity and business restoration

5. Determine Recovery Timeframes and Minimum Resource Requirements

   a. Determine recovery windows for critical business functions based on level of criticality

   b. Determine the order of recovery for critical business functions and support functions and systems based on parallel and interdependent activities

   c. Determine minimum resource requirements for recovery and resumption of critical functions and support systems

      (1)   Internal and external resources

      (2)   Owned versus nonowned resources

      (3)   Existing resources and additional resources required

6. Identify and Prioritize Business Processes

   a. Interdependencies between the business processes

        b. Process and technology dependencies
          (1) Intradepartment
          (2) Interdepartment
          (3) External relationships
  7. Determine Replacement Times
        a. Equipment
        b. Key personnel

SUBJECT AREA 4: DEVELOPING BUSINESS CONTINUITY MANAGEMENT STRATEGIES

Determine and guide the selection of alternative business recovery operating strategies for recovery of business and information technologies within the recovery time objective, while maintaining the organization's critical functions.

  A. The Professional's Role Is to:
    1. Understand Available Alternatives and Their Advantages, Disadvantages, and Cost Ranges, Including Mitigation as a Recovery Strategy
    2. Identify Viable Recovery Strategies within Business Functional Areas
    3. Consolidate Strategies
    4. Identify Off-Site Requirements and Alternative Facilities
    5. Develop Business Unit Strategies
    6. Obtain Commitment from Management for Developed Strategies
  B. The Professional Should Demonstrate a Working Knowledge in the Following Areas:
    1. Identify Enterprise-wide and Business Unit Continuity Strategic Requirements
      a. Review business continuity issues
        (1) Time frames
        (2) Options
        (3) Location
        (4) Personnel
        (5) Communications (crisis/media and voice/data)
      b. Review technology continuity issues for each support service
      c. Review nontechnology continuity issues for each support service, including those support services not dependent on technology
      d. Compare internal/external solutions
      e. Identify alternative continuity strategies
        (1) Do nothing
        (2) Defer action
        (3) Manual procedures
        (4) Reciprocal agreements
        (5) Alternative site or business facility

        (6)   Alternate source of product
        (7)   Third-party service providers/outsourcers
        (8)   Distributed processing
        (9)   Alternative communications
      (10)   Mitigation
      (11)   Preplanning
   f. Compare internal and external solutions
   g. Assess risk associated with each optional continuity strategy

2. Assess Suitability of Alternative Strategies Against the Results of a BIA
   a. Effectively analyze business needs criteria
   b. Clearly define continuity planning objectives
   c. Develop a consistent method for evaluation
   d. Set baseline criteria for continuity strategy options

3. Prepare Cost/Benefit Analysis of Continuity Strategies and Present Findings to Senior Management
   a. Employ a practical, understandable methodology
   b. Set realistic time schedules for evaluation and report writing
   c. Deliver concise, specific recommendations to senior management

4. Select Alternate Site(s) and Off-Site Storage
   a. Criteria
   b. Communications
   c. Agreement considerations
   d. Comparison techniques
   e. Acquisition
   f. Contractual consideration

5. Understand Contractual Agreements for Business Continuity Services
   a. Understand and prepare requirements statements for use in formal agreements for the provision of continuity services including jurisdictional/regulatory requirements as appropriate
   b. Formulate any necessary technical specifications for use in "invitation-to-tender" format
   c. Interpret external agreements proposed by suppliers in relation to the original requirements specified
   d. Identify specific requirements excluded from any standard agreements proposed
   e. Understand and advise on the inclusion of optional elements and those that are essential

SUBJECT AREA 5: EMERGENCY RESPONSE AND OPERATIONS

Develop and implement procedures for response and stabilizing the situation following an incident or event, including establishing and managing an Emergency Operations

Center to be used as a command center during the emergency.

  A. The Professional's Role Is to:
   1. Identify Potential Types of Emergencies and the Responses Needed (e.g., fire, hazardous materials leak, medical)
   2. Identify the Existence of Appropriate Emergency Response Procedures
   3. Recommend the Development of Emergency Procedures Where None Exist
   4. Integrate Disaster Recovery/Business Continuity Procedures with Emergency Response Procedures and Escalation Procedures
   5. Identify the Command and Control Requirements of Managing an Emergency
   6. Recommend the Development of Command and Control Procedures to Define Roles, Authority, and Communications Processes for Managing an Emergency
   7. Ensure Emergency Response Procedures are Integrated with Requirements of Public Authorities (Refer also to Subject Area 10, Coordination with Public Authorities)
  B. The Professional Should Demonstrate a Working Knowledge in the Following Areas:
   1. Identify Components of Emergency Response Procedure
     a. Reporting procedures
       (1) Internal (escalation procedures)
         (a) Local
         (b) Organization (decision-making process)
       (2) External (response procedures)
         (a) Public agencies and media
         (b) Suppliers of products and services
     b. Preincident preparation
       (1) By types of disaster
         (a) Acts of nature
         (b) Accidental
         (c) Intentional
       (2) Management continuity and authority
       (3) Roles of designated personnel
     c. Emergency actions
       (1) Evacuation
       (2) Medical care and personnel counseling
       (3) Hazardous material response
       (4) Fire fighting
       (5) Notification
       (6) Other
     d. Facility stabilization
     e. Damage mitigation
     f. Testing procedures and responsibilities

2. Develop Detailed Emergency Response Procedures
   a. Protection of personnel
      (1) Personnel assembly locations and process for ensuring identification and safety of all employees, including appropriate escalation procedures as required
      (2) Recognize and understand the value of supplementing any relevant statutory precautions
      (3) Identify options for immediate deployment and subsequent contract
      (4) Provide for communication with staff, next of kin, and dependents
      (5) Understand implications of statutory regulations
   b. Containment of incident
      (1) Understand the principles of salvage and loss containment
      (2) Understand options available to supplement the efforts of the emergency services in limiting business impact
      (3) Understand possibilities within business functions to limit the impact of a disaster
   c. Assessment of effect
      (1) Analyze the situation and provide effective assessment report
      (2) Estimate the event's direct impact on the organization
      (3) Communicate situation to employees at involved facility and any other organization locations
      (4) Demonstrate awareness of the likely media interest and formulate a response in conjunction with any existing public relations and/or existing marketing unit
   d. Decide optimum actions
      (1) Understand the issues to be considered when recommending or making decisions on continuity options
      (2) Understand the roles of the emergency services
      (3) Maintain principles of security (personnel, physical, and information)
3. Identify Command and Control Requirements
   a. Designing and equipping the Emergency Operations Center
   b. Command and decision authority roles during the incident
   c. Communication vehicles (e.g., email, radio, messengers, and cellular telephones, etc.)
   d. Logging and documentation methods
4. Command and Control Procedures
   a. Opening the Emergency Operations Center
   b. Security for the Emergency Operations Center
   c. Scheduling the Emergency Operations Center teams
   d. Management and operations of the Emergency Operations Center
   e. Closing the Emergency Operations Center

    5. Emergency Response and Triage
      a. Develop, implement, and exercise emergency response and triage procedures, including determination of priorities for actions in an emergency
      b. Develop, implement, and exercise triage procedures such as first aid and medical treatment; identify location and develop procedures for transportation to nearby hospitals
    6. Salvage and Restoration
      a. Assemble appropriate team(s)
        (1) Understand the need for effective diagnosis of incident by telephone
        (2) Understand the need for effective assembly of relevant resources at the affected site
        (3) Develop internal escalation procedures to provide required level of resources on site as incident/response develops
      b. Define strategy for initial on-site activity
        (1) Understand the need to identify immediate loss mitigation and salvage requirements
        (2) Understand the need for and, if necessary, prepare an action plan for site safety, security, and stabilization
        (3) Identify appropriate methods for protection of assets on site, including equipment, premises, and documentation
        (4) Recognize potential need to establish liaison with external agencies (e.g., statutory agencies, emergency services such as fire departments and police, insurers, loss adjusters, etc.), and specify type of information these agencies may require
        (5) Understand business requirements and interpret them to aid physical asset recovery
        (6) Establish procedures with public authorities for facility access
        (7) Establish procedures with third-party service providers, including appropriate contractual agreements

SUBJECT AREA 6: DEVELOPING AND IMPLEMENTING BUSINESS CONTINUITY AND CRISIS MANAGEMENT PLANS

Design, develop, and implement Business Continuity and Crisis Management plans that provide continuity within the recovery time objective and recovery point objective.

    A. The Professional's Role Is to:
      1. Identify the Components of the Planning Process
        a. Planning methodology
        b. Plan organization

        c. Direction of efforts

        d. Staffing requirements

   2. Control the Planning Process and Produce the Plan

   3. Implement the Plan

   4. Test the Plan

   5. Maintain the Plan

B. The Professional Should Demonstrate a Working Knowledge in the Following Areas:

   1. Determine Plan Development Requirements

        a. Roles and responsibilities

        b. Develop action plans/checklists

        c. Review and evaluate tools, e.g., business continuity planning software

        d. Acquire business processes and technology matrices and flowcharts

        e. Develop forms to acquire information

        f. Determine requirements for information database

        g. Identify other supporting documentation

   2. Define Continuity Management and Control Requirements

        a. Define scope

           (1) Identify incidents/events process may be utilized for

           (2) Suggest severity criteria that may be used to create a definition

           (3) Design escalation criteria

        b. Identify and agree on approach to key phases for continuity; document agreed approach

        c. Establish procedure to transition from emergency response plan to crisis management and/or business continuity plans

   3. Identify and Define the Format and Structure of Major Plan Components

        a. Plan designs and structures

           (1) Define how plan structures are tied to the organization

           (2) Document structure and design of plans

           (3) Ensure built-in mechanisms to ease maintenance

           (4) Define the process for gathering data required for plan completion

        b. Allocate tasks and responsibilities

           (1) Identify tasks to be undertaken

           (2) Identify necessary teams to perform required tasks

           (3) Assign responsibilities to teams

           (4) Identify and list key contacts, suppliers, and resources

   4. Draft the Plans

        a. Select appropriate tools for plan development and maintenance

        b. Draft the plans, ensuring adequate and appropriate involvement of personnel required to implement the plan

        c. Continue gathering data as needed to ensure BCP is complete and accurate

5. Define Business Continuity and Crisis Management Procedures
    a. Locate and catalogue organization information
        (1) Identify and confirm processing and documentation critical to the organization's key business
        (2) Identify and determine which information/processes should be replicated
        (3) Identify storage requirements
        (4) Identify key suppliers
        (5) Select or recommend appropriate methods of business backup including understanding of retention periods and duplication/replication schedules, etc.
    b. Information continuity
        (1) Recommend and develop appropriate procedures, taking into account:
            (a) Business requirements
            (b) Technology requirements
            (c) Legislative requirements
    c. Process continuity
        (1) Recommend alternative ways to conduct when normal resources are available following a disaster or other disruptive event that will be effective until continuity procedures are successfully implemented
        (2) Recommend method/procedures to easily transfer business functions from any alternative, temporary, or emergency operation into the new, replaced, or reinstalled service
        (3) Identify critical equipment; acquisition and/or reconditioning mainframes
6. Damage Assessment/Restoration Strategy
    a. Damage assessment
        (1) Create an action plan for assessing damage, including:
            (a) Understand economics of repair versus replacement
            (b) Understand the capabilities of salvage specialists inselecting and applying relevant methods of contamination analysis
            (c) Understand the criteria for selecting appropriate subcontractors for salvage operations
        (2) Clearly relate damage assessment to business continuity of organization
    b. Define restoration strategy
        (1) Employ a logical, but relevant and practical, approach to business recovery requirements
        (2) Demonstrate ability to reduce consequential losses
        (3) Agree upon restoration methods for business assets (e.g., equipment, electronics, documents, data, furnishings, premises, plant, computers, etc.)
        (4) Understand the approval process for restoration and, especially, the implications of warranties
        (5) Define a strategy for restoration

7. Develop General Introduction or Overview
   a. General information
      (1) Introduction
      (2) Scope
      (3) Objectives
      (4) Assumptions
      (5) Responsibility overview
      (6) Testing
      (7) Maintenance
   b. Plan activation
      (1) Notification
         (a) Primary
         (b) Secondary
      (2) Disaster declaration procedures
      (3) Mobilization procedures
      (4) Damage assessment concepts
         (a) Initial
         (b) Detailed
         (c) Team members
   c. Team organization
      (1) Team description
      (2) Team organization
      (3) Team leader responsibilities
   d. Policy statement
   e. Emergency Operations Center
8. Develop Administration Team Documentation
   a. Identify continuity functions for the following, including qualifications, responsibilities, and resources required
      (1) Communications (public relations/media, client, and employee)
      (2) Personnel/human resources
      (3) Security
      (4) Insurance/risk management
      (5) Equipment/supplies purchasing
      (6) Transportation
      (7) Legal
   b. Other specialist coordinator/team responsibilities
      (1) Relations/liaison with regulatory bodies
      (2) Investor relations
      (3) Relations with other involved groups (e.g., customers and suppliers)
      (4) Labor relations
   c. Develop specific procedures for each function or building identified above:
      (1) Department/individual/building plans
      (2) Checklists
      (3) Technical procedures

9. Develop Business Operations Team Documentation
   a. Operating department plans
      (1) Essential business functions
      (2) Information protection and recovery
      (3) Activation actions
      (4) Disaster site recovery/restoration actions
      (5) End-user computing needs
   b. Action sections
      (1) Recovery team
         (a) Personnel
         (b) Responsibilities
         (c) Resources
   c. Action plans
      (1) Specific department/individual plans
      (2) Checklists
      (3) Technical procedures
10. Develop Information Technology Recovery Team Documentation
   a. Recovery site activation
      (1) Management
      (2) Administration/logistics
      (3) New equipment
      (4) Technical services
      (5) Application support
      (6) Network communications
      (7) Network engineering
      (8) Operations
      (9) Intersite logistics and communications
      (10) Data preparation
      (11) Production control
      (12) End-user liaison
   b. End-user requirements
   c. Identify components of vital records program
   d. Action sections
      (1) Recovery team
         (a) Personnel
         (b) Responsibilities
         (c) Resources
   e. Action plans
      (1) Specific department/individual plans
      (2) Checklists
      (3) Technical procedures
11. Develop Communication Systems
   a. Voice communications recovery plans
      (1) Phone lines, including in-bound, toll-free (1-800) lines, and fax lines

          (2) Voice mail, voice response units, and other voice-based services

          (3) Alternate arrangement for automated voice response during a disaster

     b. Data communications recovery plans

          (1) Data communications with mainframe-based information systems

          (2) Local area network (LAN) recovery for work area recovery

          (3) Wide area network (WAN) recovery for restoring global connectivity

          (4) Email, groupware, and other data communications–based work support

     c. Emphasize and ensure detailed and up-to-date documentation of voice and data communications networks throughout the enterprise

12. Develop End-User Applications Plans

     a. Plan design and structure

          (1) Identify examples of alternative plans and structures

          (2) Define how plan structure is tied to the organization

          (3) Document structure and design of departmental continuity plans

          (4) Ensure built-in mechanisms to ease maintenance

          (5) Plan and implement the gathering of data required for plan completion

     b. Identify and agree on approach to key phases of recovery; document agreed approach

     c. Allocate tasks and responsibilities

          (1) Differentiate between recovery teams and departmental teams

          (2) Identify tasks to be undertaken

          (3) Identify necessary teams to perform required tasks

          (4) Assign responsibilities to teams

          (5) Identify and list key contacts, suppliers, and resources

13. Implement the Plans

     a. Ensure that required tasks are completed for plan implementation

          (1) Acquiring additional equipment

          (2) Contractual arrangements

          (3) Preparing backup and off-site storage

          (4) Appropriate documentation for plans in place

     b. Develop test plans, schedules, and test reporting procedures

          (1) Acquiring additional equipment

          (2) Contractual arrangements

          (3) Preparing backup and off-site storage

c. Develop maintenance, updating, and reporting procedures
14. Establish Plan Distribution and Control Procedures
   a. Establish procedures for distribution and control of business continuity plans
   b. Establish procedures for distribution and control of results of plan exercises
   c. Establish procedures for distribution and control of plan changes and updates

SUBJECT AREA 7: AWARENESS AND TRAINING PROGRAMS

Prepare a program to create and maintain corporate awareness and enhance the skills required to develop and implement the Business Continuity Management program or process and its supporting activities.

A. The Professional's Role is to:
   1. Establish Objectives and Components of Corporate BCM Awareness and Training Program
   2. Identify Functional Awareness and Training Requirements
   3. Develop Awareness and Training Methodology
   4. Acquire or Develop Awareness and Training Tools
   5. Identify External Awareness and Training Opportunities
   6. Identify Alternative Options for Corporate Awareness and Training
B. The Professional Should Demonstrate a Working Knowledge in the Following Areas:
   1. Define Awareness and Training Objectives
   2. Develop and Deliver Various Types of Training Programs as appropriate
      a. Computer-based
      b. Classroom
      c. Test-based
      d. Instructional guides and templates
   3. Develop Awareness Programs
      a. Management
      b. Team members
      c. New employee orientation and current employee refresher program
   4. Identify Other Opportunities for Education
      a. Professional business continuity planning conferences and seminars
      b. User groups and associations
      c. Publications and related Internet sites

Subject Area 8: Maintaining and Exercising Business Continuity Plans

Preplan and coordinate plan exercises, and evaluate and document plan exercise results. Develop processes to maintain the currency of continuity capabilities and the plan documents in accordance with the organization's strategic direction. Verify that the plans will prove effective by comparison with a suitable standard, and report results in a clear and concise manner.

A. The Professional's Role is to:
1. Preplan and Coordinate the Exercises
2. Facilitate the Exercises
3. Evaluate and Document the Exercise Results
4. Update the Plan
5. Report Results/Evaluation to Management
6. Coordinate Ongoing Plan Maintenance
7. Assist in Establishing Audit Program for the Business Continuity Plan
B. The Professional Should Demonstrate a Working Knowledge in the Following Areas:
1. Establish an Exercise Program
   a. Develop an exercise strategy that does not put the organization at risk; is practical, cost-effective, and appropriate to the organization; and ensures a high level of confidence in recovery capability
   b. Employ a logical, structured approach (effectively analyze complex issues)
   c. Create a suitable set of exercise guidelines
2. Determine Exercise Requirements
   a. Define exercise objectives and establish acceptable levels of success
   b. Identify types of exercises and their advantages and disadvantages
      (1) Walk-throughs/tabletop
      (2) Simulations
      (3) Modular/component (call trees, applications, etc.)
      (4) Functional (specific lines of business)
      (5) Announced/planned
      (6) Unannounced/surprised
   c. Establish and document scope of the exercise (participants, timing, etc.)
3. Develop Realistic Scenarios
   a. Create exercise scenarios to approximate the types of incidents the organization is likely to experience and the problems associated with these incidents
   b. Map scenarios identified to different test types

4. Establish Exercise Evaluation Criteria and Document Findings
   a. Develop criteria aligned with exercise objectives and scope
      (1) Measurable and quantitative
      (2) Qualitative
   b. Document results as per criteria identified
      (1) Expected versus actual results
      (2) Unexpected results
5. Create an Exercise Schedule
   a. Develop a progressive, incremental schedule
   b. Set realistic time scales
6. Prepare Exercise Control Plan and Reports
   a. Define exercise objectives and select an appropriate scenario
   b. Define assumptions and describe limitations
   c. Identify resources required to conduct the exercise, identify participants; ensure all understand the objectives and their roles
   d. Identity exercise adjudicators (umpires), and clearly identify all roles and responsibilities
   e. Provide an inventory of items required for the exercise and specifications for the exercise environment
   f. Provide a timetable of events and circulate to all participants, facilitators, and adjudicators
   g. In the event of a real situation occurring during an exercise, you may want to have a predetermined mechanism for cancelling the exercise and invoking your real business continuity process
7. Facilitate Exercises
   a. Execute the exercise(s) as planned above
   b. Audit exercise actions
8. Post-Exercise Reporting
   a. Provide a cogent, comprehensive summary with recommendations, commensurate with levels of confidentiality requested by exercise umpire/adjudicator or as specified by the subject organization
9. Feedback and Monitor Actions Resulting from Exercise
   a. Conduct debriefing sessions to review exercise results and identify action items for improvement
   b. Identify actions and owners for recommendations; confirm owner acceptance
   c. Confirm time schedules for completing or reviewing agreed actions
   d. Monitor (and escalate where necessary) progress to completion of agreed actions
10. Define Plan Maintenance Scheme and Schedule
    a. Define ownership of plan data
    b. Prepare maintenance schedules and review procedures
       (1) Select tools
       (2) Monitor activities

     (3) Establish update process

     (4) Audit and control

   c. Ensure that scheduled plan maintenance addresses all documented recommendations

  11. Formulate Change Control Procedures

   a. Analyze business changes with business continuity planning implications

   b. Set guidelines for feedback of changes to planning function

   c. Develop change control procedures to monitor changes

   d. Create proper version control—develop plan reissue, distribution, and circulation procedures

   e. Identify plan distribution list for circulation

  12. Establish Status Reporting Procedures

   a. Content

   b. Frequency

   c. Recipients

  13. Audit Objectives

   a. Recommend and agree on objectives for BCM-related audits

   b. Audit the BCP's Structure, Contents, and Action Sections

    (1) Determine whether a section in the BCP addresses recovery considerations

    (2) Evaluate the adequacy of emergency provisions and procedures

    (3) Recommend improved positions if weaknesses exist

   c. Audit the BCP's Documentation Control Procedures

    (1) Determine whether the BCP is available to key personnel

    (2) Review update procedures

    (3) Demonstrate that update procedures are effective

    (4) Examine the provision of secure backup copies of the BCP for emergency use

    (5) List those individuals with copies of the BCP

    (6) Ensure that BCP copies are current

SUBJECT AREA 9: PUBLIC RELATIONS AND CRISIS COMMUNICATIONS

Develop, coordinate, evaluate, and exercise plans to communicate with internal stakeholders (employees, corporate management, etc.), external stakeholders (customers, shareholders, vendors, suppliers, etc.), and the media (print, radio, television, Internet, etc.).

  A. The Professional's Role is to:

   1. Establish Programs for Proactive Crisis Communications

Sample Business Impact Analysis Introduction Letter

January 1, 2006

Iva Bucks
Chief Financial Officer
XYZ Corporation
4100 Enterprise Drive
Palo Alto, CA 94025

Dear Mr. Bucks:

XYZ Corporation asked the security department to assist the company with the development of business continuity (disaster recovery) plans for its major facilities at Menlo Park and Dublin. This request was motivated by concerns about regulatory requirements, the potential for loss from an extended business interruption, and the degree of similar planning by major competitors. Also, the financial auditors listed the lack of effective recovery plans as a deficiency on their report.

The basis or starting point for this process is the development of a Business Impact Analysis. During our meeting scheduled for Tuesday, April 4, 2006, at 10:00 A.M., we will ask you to help us determine the financial and subjective impact to XYZ from the loss of individual business functions over a period of time. We will use this information to:

- Demonstrate loss potential to senior management
- Form the basis for the business continuity planning process
- Evaluate or verify a function's "outage tolerance (Recovery Time Objective)"
- Prioritize recovery actions and resources
- Focus our questions to other managers we may need to interview

We will provide a questionnaire to help explain what information is needed and to assist you and your staff with the assembly of this information. All data and results are maintained according to our confidentiality agreements.

Sincerely,
Eugene Tucker, CPP, CFE, CBCP
Assistant Vice President
Business Continuity Planning

Sample Kidnap
and Ransom Contingency Plan

This plan was designed to fit the needs of a particular company with a unique kidnap and ransom exposure. It is not a blueprint that can or should be followed by any other organization, although parts of it will apply to most kidnap and ransom situations.

I. INTRODUCTION

Even the most carefully tailored security procedures sometimes are not enough. As evidenced by repeated bombings, kidnappings, extortion plots, and other acts of violence, security procedures are not infallible; they can be compromised or penetrated. For this reason, it is vitally important for this company to have a written contingency plan that outlines some of the steps to be taken in the event a kidnap or extortion plot against the company becomes an actuality. The threat/risk ratio for _____ has been assessed as low to medium. Therefore, some security precautions are deemed to be essential. Should the threat/risk ratio change, it will be necessary to review the existing security program to reassess the level of security and determine whether it is adequate in view of the changing circumstances.

    Note: The contents of this report are confidential. The details contained herein should receive limited distribution. The plan is classified as "Business Confidential."

II. BASIC PLAN

A. Policy Regarding Ransom

Company policy is that a reasonable ransom or extortion will be paid in the event of a kidnap of one or more corporate officials or members of their immediate families. The same applies in the event of an extortionate plot against the company. The limitations of our insurance coverage are as follows:

        1. Dollar limitations
        2. People covered (general)

    Note: A copy of the insurance policy covering kidnap and extortion should be attached hereto as an exhibit and remain a permanent part of this file. [The existence of this document is always confidential.]

B. Crisis Management Team Composition

Members of the crisis management team should include those few individuals having authority to implement and carry out the policy as dictated by the board of directors and the procedures contained in this plan. The presence of more than five people on this team could easily lead to confusion at a time when confusion is least desirable. In addition, the team should be aware that all the resources, in terms of manpower and material, of the company are available for their use on an ad hoc basis.

Members of the team should include:

1. The coordinator, chairman of the board, or chief executive officer
2. President and chief operating officer
3. Executive vice president of finance
4. Executive vice president of operations
5. Executive vice president of administrative and technical services

Other members of the team may vary, depending on the nature of the threat or demand and whether the crisis occurs in [company's location], the United States, or a foreign country.

C. The Coordinator (and Alternate)

The coordinator's function is to implement the plan and procedures and to coordinate the crisis operation. The coordinator should also be the person with the top decision-making authority. In the event that _____ becomes a victim, the next person in line of succession in the crisis management team would be the alternate coordinator. In the event he or she is unavailable, the next in line of succession would be_____. [The line of succession must be worked out in advance. You may wish to reduce this succession policy to writing. A copy of that policy should then be attached to this document as an exhibit.]

The coordinator, _____ (code name "Mr./Ms. Adams," when dealing with a kidnaper or extortionist) will not act as the negotiator. It is vitally important that the task of negotiator be assigned to only one person—a person who has had training as a negotiator in criminal situations. In this case, we strongly recommend that _____, Vice President, Industrial Relations, who has been trained in union bargaining and negotiations, be trained for criminal-type negotiations. In the event of a kidnap or ransom against the company, [he/she] would become an ad hoc member of the team, serving in the capacity as negotiator and advisor. Mr./Ms. _____ should also direct the preparation of the list of names and related information, which will become a permanent part of this plan (see Exhibit E.1).

In addition, the coordinator shall:

1. Formulate plans and procedures for handling crisis situations.
2. Gather an advisory staff (if deemed appropriate) to generate information and perform services to facilitate these procedures.
   Example: A member of the legal staff may be necessary to review the plans for compliance with established corporate policy.

3. Maintain in a secure place the current crisis management plan and procedures.

4. Communicate these plans and procedures to only authorized individuals, and follow-up to ensure that these individuals are fully cognizant of any changes in the plan or procedures.

5. Maintain current personal information and biographies of all corporate executives in a secure place. The personnel department maintains a very limited amount of biographical information pertaining to company executives. Enclosed with this document is a biographical inventory. We recommend all executives complete the document, to be placed in their individual personnel files where they can be quickly located in the event of an emergency.

6. Recruit and train the personnel necessary to carry out the crisis management program.

7. Exercise good judgment in determining the course of action in any crisis situation not covered by approved policy.

8. Implement plans and procedures according to the management plan.

Exhibit E.1.   List of Executives and Personnel

1. Executives and Publicity Identified Personnel

Name	Title
_____	_____
_____	_____

2. Branch or Profit Center Executive Personnel

Name	Title
_____	_____
_____	_____

3. Personnel to Authorize Ransom Payment (2 needed)

_____

_____

4. Name of Financial Institution

Contact: _____

Telephone Number: _____

5. Personnel to Administer Payment and Plan/Draw Payment (1)

_____

7. Personnel to Handle Police, Press Contact (1)

Press: _____

Police: _____

Federal Bureau of Investigation: _____

8. Persons to Be Notified of Demand

Name	Telephone (Office)	(Home)
_____	_____	_____
_____	_____	_____
_____	_____	_____

D. The Crisis Management Center

The purpose of the crisis management center is to serve as the focal point for directing a coordinated and planned response during a crisis situation. It should be located within the organization's headquarters facility at or in the executive conference room. It should be furnished with all documents, supplies, and communications that may be needed during a crisis. At the minimum, items such as tape recorders, office equipment, computers, and a log to record all calls and actions taken will be necessary.

E. Crisis Management Plan (CMP) Implementation

When an executive, employee, or family member becomes the victim of a kidnap, or the company becomes the victim of an extortion or terrorist plot, the organization will respond by implementing the CMP. The authority to implement the plan must be clearly spelled out. Implementation criteria should be defined:

1. Who has the authority to implement the crisis management plan?
   a. Chairman of the board or chief executive officer
   b. Alternatively, the president and chief operating officer
2. What are the minimum circumstances that must exist for this authority to become effective (example: if a threat of kidnapping is received)?
   a. Time period of duration that this authority will remain in effect is: (example: until the crisis is satisfactorily resolved)
   b. Succession of this authority should the holder be removed or incapacitated: (names)

F. Crisis Management Program

When an extortion demand or threat is received, it should be immediately reported to the decision-making authority, as outlined in the crisis management plan. The decision as to when this crisis management program should be implemented will depend on the following factors:

1. Threat verification (true/false)
2. Threat analysis:
   a. How valid is the threat?
   b. Who is doing the threatening?
      • Terrorists (political). Like their counterparts in other countries, these people are the most dangerous. One of their principal goals is publicity, which can be accomplished most effectively by shock tactics. A large multinational company or utility represents a prime target.
      • Criminals. Many criminals in foreign countries have turned to kidnapping, extortion, and other terrorist tactics for criminal

gain. Because the goals of these people are identifiable, they are usually open to bargaining, and their demands can be negotiated.

- The Mentally Ill. This is the fanatic whose sense of values is at odds with those of society. This person if often prepared to die or go to jail for a cause. In this category fall the cunning, the clever, and the inept. They are always unpredictable and therefore hard to plan for and deal with. Also, as in the case of the "Unabomber," they are also difficult to identify, locate, and arrest.

Note: Most kidnappings are carried out by criminals or the mentally ill. Although kidnappers seeking publicity or nonmonetary rewards usually select large companies, those wanting money frequently select a prominent official of a medium-sized company. Most such organizations have considerable amounts of money available and do not have blanket policies against paying ransom money.

Note: Threat analysis in complicated cases is usually best left to trained security, risk management, or law enforcement personnel. The tools of threat analysis should be used only as aids in decision making. Total reliance on any one method or tool may cause serious error. In a situation as complicated and dangerous as a kidnapping, all factors must be weighed in order to arrive at effective resolutions.

G. Verification of the Validity of the Threat

1. Does a threat exist?
2. Is the threat as serious, more serious, or less serious than the creator of the threat would have us believe?
3. What is the present vulnerability of the intended victim of the threat? How will this vulnerability increase if we:
   a. Ignore the demands?
   b. Grant the demands?
   c. Engage in prolonged negotiations?

H. The Threateners

1. Can we identify the individual or group responsible for the threat?
2. Can we pinpoint the origin of the threat (physical location)?
3. What is the previous history or experience of this type of threat in this specific environment? (Law enforcement input here is usually necessary.)
4. What is the previous history of other organizations experiencing this type of threat in this specific environment? (In some foreign locations, the police may not be privy to these data.)
5. What type of threat are we faced with?
   a. A simple extortion?
   b. A simple threat, no demands?

      c. A demand without a threat?

      d. What does the type of threat indicate about the demanders' view of the company?

      e. What does the type of threat indicate about the group or person making the threat?

  6. How was the threat delivered?

      a. Verbally, telephone (see Exhibit E.2, ransom demand telephone checklist)

Exhibit E.2. Ransom Demand Telephone Checklist

Time of Call _____

Make every attempt to gain as much information from the caller as he will furnish, but do not give the caller the impression you are reading questions from a checklist or are trying to keep him on the line so the call can be traced. Write down the responses of the caller *word for word*.

_____

Would you please repeat your statement?

_____

Who is making this demand?

_____

How do I know this is not a joke? We get many pranks here.

_____

IF A KIDNAP:

What is (he, she) wearing? _____

May I talk to (him, her)? _____

Could you explain what you want? _____

I will have to give your demands to my superior. We will want you to include the word (key word)* and the number (key number)** in all future communications with us.

If the caller gets into specifics on payment, ask, "What do you want"?

_____

If money: what currency and how do you want it? _____

Where and when should the ransom be delivered? _____

How should the payment be made? _____

End the call on a positive note, by assuring the caller his demand will be communicated to the proper person in the company as soon as possible. Leave the caller with the impression that his or her call has been understood and action will be taken. Make note of the following information.

Time call ended _____

Background noises _____

Sex of caller _____

Approximate age _____

Any accent _____

Any voice peculiarity such as lisp or stutter _____

What was the caller's attitude? _____

Was the caller sober? _____

Did the caller sound educated? _____

What did you notice about the call that you find unusual? _____

If the caller seemed familiar with the building or operation indicate how _____

*Recognition code has been established as: Mr. Adams.

**Private unlisted telephone number.

   b. By messenger:
      i. Delivered
      ii. Mailed
      iii. Found at crime scene in a protected area
   c. What does the manner in which the threat was delivered indicate about:
      i. The individual or group making the threat?
      ii. The location of the individual or the group making the threat?
      iii. The nature of the group or individual making the threat?
7. What is the nature of the demand, if any?
   a. Who, or what, is the precise victim?
   b. Who is or will become the victim(s) if the demand is not met?
   c. Are demands:
      i. Within the realm of possibility? (Release of political prisoners, for example, is something most corporations cannot influence, much less accomplish.)
      ii. Of propaganda value to the threatener? (Usually a tactic of terrorists)
8. Who is making the demand?

Note: All the above information plus any other data available that may be helpful must be collected and analyzed to evaluate the validity of a threat. Time is a critical element in threat analysis. In situations in which time is very short, it must not be wasted on deciding what categories of data are most necessary to be collected before analysis.

All genuine threats will be manifestations of careful preplanning. Only a preplanned response will suffice to meet such a threat. Such preplanning in threat analysis must begin at the very inception of the threat. Remember, time is of the essence. Few sophisticated kidnappers (extortionists) will allow you much time for decision making. Much of the necessary planning can occur before the receipt of a threat.

I. Resources of Verification

Listed below are some resources one might consider using to verify the genuineness of a threat, providing of course that there is sufficient time:

1. Corporation-processed pre-event data (may indicate a kidnapper has inside information)
2. Prearranged codes and procedures. For example, "Mr./Ms. Adams" (the coordinator) to "Mr./Ms. Able" (the kidnapper-extortionist). This will preclude an opportunist from taking advantage of publicity to divert an extortion payment or otherwise interfere with the recovery.
3. Local and national law enforcement liaison
   a. The Federal Bureau of Investigation (FBI)
   b. The Department of Public Safety (state level)
4. The Office of Security, U.S. Department of State (if the victim is overseas)

5. Host government liaison (if overseas)
6. Propaganda analysis (for terrorists)
7. Psychological stress evaluator (PSE), if conversation tapes are available
8. Psychiatric analysis
9. Psycholinguistics (tapes only)
10. Graphology (document examination)
11. Forensic document examination
12. Voice-print analysis (tapes only)
13. Electronic tracing
14. Noise analysis
15. Previous case histories

Note: In most cases, the initial threat is communicated by telephone, so that demands can receive maximum attention with a minimum of delay. The information contained in the original threat messages is of vital importance.

Ideally, the initial threat message should be recorded. In practice, this is seldom accomplished, although, with preplanning, any secondary and beyond-threat messages can and should be recorded. (Usually the FBI or local law enforcement will arrange to do this.)

Alternatively, the threat call must be reduced to detailed notes. A form to assist in this task is attached and should be furnished to the central telephone operator, or the person who usually handles incoming calls. It should be located at or near the telephone in an inconspicuous place, for ready reference (see Exhibit E.2, sample ransom demand telephone checklist).

All demands must be communicated to the crisis management team by oral or written message. Analysis of the communication itself can often reveal a great deal about the character of the individual making the demand and may also reveal whether the threat implicit in the demand is genuine.

J. Analysis

The purpose of threat analysis is to turn any form of threat into a manageable problem that can be analyzed and then eliminated, neutralized, or controlled by a crisis management team. A schematic of the crisis management process is as follows:

1. Preplanning
   a. Resource identification
   b. Crisis program operating on standby
2. Threat reception: competent reception of threat and the circumstances of its receipt
3. Threat verification: what must we know to be certain that the threat is real?
4. Threat analysis: what must we know to determine:
   a. Threat level?
   b. Identification of the threateners?
   c. Safety of personnel/assets?
   d. Validity of negotiations?

        e. Origin of threat and location of our personnel?

        f. Real goal of the threatener?

        g. Creation of a risk matrix (if deemed necessary)?

    5. Threat response: what steps must be taken to eliminate, neutralize, or control the threat and guarantee the safety of our personnel and assets?

K. Extortion Demands

The demands criminals, terrorists, and mentally ill individuals make in these cases usually fall into one or more of the following categories:

    1. The amount of ransom money for the safe return of a kidnapped executive or a member of his or her family depends on:

        a. The wealth of the organization

        b. The criminal's (terrorist's) needs

        c. An intention to demand a "measured quota" from the organization

        d. "Value" of the kidnap victim in the eyes of the criminal (perception)

    2. Medical supplies for hospitals, public works, and the like, in exchange for the hostage

    3. Public recognition of their cause (terrorists)

    4. Release of fellow terrorists or members of their organization jailed by authorities

    5. Protest against national politics and policies, or those of the victim organization

    6. To embarrass the organization, victim, or victim's family

Note: Prenegotiation preparation and training should cover the possibility of more than one of the above demands being presented.

L. Presentation of Demands

Characteristics of the demands of some criminals and most terrorist organizations are:

    1. The demands are nonnegotiable (at least in the beginning).

    2. All demands (if more than one) must be met in full.

    3. Time periods are usually short and are often established at the outset. The initial demands and timeframes are rarely realistic.

    4. The consequences may be the prompt carrying out of the threat if the demands are not met.

M. Response

The response of the organization to the demand will probably be determined by the policy of agreeing to negotiate for "reasonable" ransom demands, as set forth above (paragraph II.A).

N. Counterdemands and Proposals

Depending on the information available, the crisis management team may respond as follows:

1. By asking for (actually, demanding) proof that the executive (victim) is still alive and unharmed. (Captors can be required to supply an item of personal identification; however, a handwritten letter—containing a key phrase or code that we dictate—would be the preferred proof. The letter should contain the date and time it was written.)
2. By asking for the exact time and place of the executive's (victim's) release if an agreement on demands can be reached.
3. By asking for time to study the demands and raise the currency. (You may not get it, but ask for it anyway.)

   Note: It is extremely important that the crisis management team signal that all reasonable demands can be negotiated and will be met, provided that the safety of the executive (victim) can be assured. The reverse should also be emphasized—that without a firm guarantee that the victim will be released unharmed, neither the money nor any other demand will be delivered. The remaining part of the plan can then be accomplished by negotiation. The best policy regarding negotiation is to play for time, total agreement, and guarantees.

O. Insurance

The insurance policy should be given close scrutiny at this point. Look closely at the coverage and restrictions to ensure that you are in full compliance. The following items should be reviewed:

1. Publicity regarding the policy
2. Genuineness of the extortion demand
3. The specific names or titles of the people covered in the policy
4. Coverage of executives in particular job categories
5. Whether payment must be made under duress
6. Cooperation by the insured with law enforcement

Exhibit E.3.   Negotiator with the Family

---

1. KEEP THE FAMILY AS CALM AND AS UNWORRIED AS POSSIBLE. Assure the family that the company is doing, and will do, everything possible for the safe release and return of the victim. Assist the family by doing small chores. Try to have the family resume normal activities as far as is possible. Act cheerful and confident.
2. If possible, the family members should not be interviewed by the news media. Younger family members, especially teenagers, might reveal detailed information that might jeopardize the success of your operation.
3. If necessary for its safety, comfort, and security, move the family to a safe haven—for instance, a motel in a secluded location—until the situation is resolved. The police will usually cooperate by assigning protection to the family. If not, hire a private bodyguard.
4. COOPERATE WITH OTHER NEGOTIATORS SO THAT EVERYONE CONCERNED IS AWARE OF THE SITUATION. Do not discuss the situation with anyone other than your fellow negotiators.

Exhibit E.4. Negotiator with Law Enforcement

1. AS SOON AS PRACTICAL, NOTIFY LAW ENFORCEMENT OF THE SITUATION. Most terrorist situations are under the joint jurisdiction of federal and local authorities. The Federal Bureau of Investigation (FBI) and the U.S. Postal Service are concerned with possible federal law violations; local police are concerned with possible local law violations.
2. BE HONEST AND FRANK WITH LAW ENFORCEMENT. In all situations, law enforcement will cooperate with you, and with each other, for the safe return of the victim. Law enforcement agencies will do nothing to jeopardize the safe return of the victim and will conduct investigations in a covert manner until the victim is returned.
3. HONESTLY ASSESS THE CAPABILITIES OF LAW ENFORCEMENT. In some communities, local police have limited capabilities and experience. In such cases, it will be better to notify the FBI first, for their primary investigative activity and to make sure that a capable investigation is conducted. Local law enforcement, in this case, would handle secondary investigation, upon notification by the FBI.
4. COOPERATE WITH OTHER NEGOTIATORS. Cooperate so that everyone concerned is aware of the situation. Do not discuss the situation with anyone other than your fellow negotiators.
5. LOCAL POLICE ARE CLOSELY CONNECTED TO THE PRESS. A telephone call to the local police switchboard or emergency number will usually be monitored by the press and television news reporters. If you are to exercise any control over the press, let the FBI notify the local police in every instance.

Exhibit E.5. Negotiator with the Media

1. REMAIN IN CONTROL OF ALL INFORMATION RELEASED TO THE MEDIA. It is better for the safe return of the victim that as little detailed information as possible be released to the press, television, and radio. You can admit that a situation exists, but do not reveal any details about the family situation, the amount of ransom demanded, or the details of the ransom delivery. In a terrorist situation where the safety of personnel is at stake, it is better to release too little rather than too much information. Law enforcement advice should be sought about the release of specific details. Criminals and terrorists read the papers and listen to news broadcasts. If too many details are furnished:
   - Sick, antisocial, or greedy people may enter the picture and complicate things by fraudulent attempts to get money (this is why code names are necessary in all negotiations with the kidnappers);
   - Overly aggressive news reporters might complicate funds delivery by close surveillance;
   - The efforts of law enforcement to secure the return of the victim, to apprehend the kidnappers, or to recover the ransom funds might be jeopardized.
2. A SINGLE PERSON SHOULD HANDLE ALL CONTACTS WITH THE MEDIA. All other people should refer any contacts from the media to that one person. This will prevent the media from playing one official against another to obtain more information. It will also permit the controlled release of nonvital information. It is standard media procedure to induce a person, by feeding his or her self-importance, to release small details that are then used to confront a second person in an attempt to extract more substantial information.
3. DO RELEASE TO THE MEDIA ALL DETAILS ABOUT SPECIAL MEDICATION OR TREATMENT NEEDED BY THE VICTIM.
4. REMEMBER AT ALL TIMES THAT YOU DO NOT HAVE TO ANSWER ANY QUESTIONS by the media or anyone else, except in a court under subpoena. Many people feel that they should, or have to, answer questions from the media or well-intentioned citizens. Learn to say, "I'll have to get back to you later with the answer to that question."
5. COOPERATE WITH OTHER NEGOTIATORS so that everyone concerned is aware of the situation. Do not discuss the situation with anyone other than your fellow negotiators.

Exhibit E.6.   Negotiator with Terrorists

1. TRY TO MAKE SURE THE VICTIM IS ALIVE. If you can talk to the terrorist, tell him honestly:
   - That you will do everything for the release of the victim.
   - That you just want to make sure that the victim is unharmed.
   - That you would like the victim to write a note to you or say something to you so that you know positively that he is alive and unharmed. (Dictate a key phrase to be included in the note.)
2. ASK THE TERRORIST TO REFER TO HIM OR HERSELF BY A CODE NAME that you agree upon, so that you will know that you are talking to the same person each time. Do not reveal the code name to anybody until after the victim is released. It is common for several different people to try to collect a ransom in any publicized kidnapping. Give each person calling a different code name. Use a neutral or even a favorable code name, not a derogatory one—"Robin Hood," rather than "Dirty Tom."
3. OBTAIN AND REPEAT INSTRUCTIONS FOR FUNDS DELIVERY. If possible, work out alternate instructions in case you cannot comply fully with the original instructions.
4. ENDEAVOR TO LESSEN THE AMOUNT OF FUNDS BEING DELIVERED. Since the safety of the victim is most important, do not haggle over the amount of funds. However, if the opportunity arises, tell the terrorist that you can obtain one-half or one-third for immediate, same-day delivery to any spot he or she wants, but that delivery of the full amount might take longer since you need higher authorization. If the terrorist is intransigent, drop the subject immediately.
5. ASSURE THE TERRORIST THAT THE TELEPHONE IS NOT BEING TAPPED, THAT LAW ENFORCEMENT HAS NOT BEEN INVOLVED, AND THAT THE NEWS MEDIA IS BEING KEPT OUT (but only if he or she brings up these subjects first).
6. COOPERATE WITH OTHER NEGOTIATORS so that everyone concerned is aware of the situation. Do not discuss the situation with anyone other than your fellow negotiators.

Exhibit E.7.   Sample Notification of Company Policy

The personal safety and well-being of all of your employees and their families is very important to the company. This cannot be overemphasized. While your company does not believe that the company, or the employees, will be the object of any criminal actions, the following procedures for action are being set out for your guidance.

In any situation involving a hostage, ransom, or extortion, the only important consideration is the safety of our personnel, of their family members, and their safe return.

In any criminal situation or in any questionable situation, notify a company official as promptly and as completely as is possible. Do not delay action to investigate the matter fully just so you can give complete details to a company official.

The company official, and his alternate, are:

Notify _____

_____ Office Phone

_____ Home Phone

_____ Alternative

_____ Office Phone

_____ Home Phone

Distribution: All company offices, managers, and supervisory personnel.

Exhibit E.8. Executive Biographical Inventory (to be retained in the individual's personnel file)

NAME: _____

NICKNAME (FOR IDENTIFICATION): _____

ADDRESS: _____

ALTERNATE ADDRESS (SUMMER, ETC.): _____

DESCRIPTION: AGE_____ BIRTHDATE_____

    PLACE OF BIRTH_____ HEIGHT_____

    WEIGHT_____ COLOR OF HAIR_____

    SEX_____ NOTICEABLE PHYSICAL TRAITS_____

HOME PHONE: _____

WIFE: _____

ADDRESS OF WIFE: _____

DESCRIPTION OF WIFE: _____

VEHICLES: _____

LICENSE	STATE	DESCRIPTION
1. _____	_____	_____
2. _____	_____	_____
3. _____	_____	_____

CODE: FAVORITE BOOK_____ FAVORITE SPORT_____

MEDICAL EMERGENCY INFORMATION: _____

DOCTOR'S NAME: _____ PHONE: _____

SPECIAL MEDICATION: _____

(BACK OF CARD)

HOUSEHOLD MEMBERS:

1.   NAME: _____ RELATIONSHIP: _____

    ADDRESS: _____

DESCRIPTION: _____

MEDICAL EMERGENCY INFORMATION: _____

2.   NAME: _____ RELATIONSHIP:_____

    ADDRESS: _____

DESCRIPTION: _____

MEDICAL EMERGENCY INFORMATION: _____

3.   NAME: _____ RELATIONSHIP:_____

    ADDRESS: _____

DESCRIPTION: _____

MEDICAL EMERGENCY INFORMATION: _____

FINGERPRINTED? _____

BLOOD TYPE: _____

LOCATION IF NOT IN FILE: _____

RECENT PHOTOGRAPH: _____

LOCATION IF NOT IN FILE: _____

How to Establish Notice

SOURCES OF DATA

Local Police Departments

Check with the crime prevention bureau, department of statistics, administration, or public information officer (PIO). Some departments use software programs to print color maps of crime incidents for selected distances around a particular address. Check with other agencies such as transit police for information on crimes committed in their jurisdictions.

News Media

Newspapers, television, and radio reporters, as well as their archives, are good sources of information. These can be researched in person at the publication's office, through a local library, or by using on-line searching.

Subpoena

This can include police records; crime prevention or physical security surveys completed by police, security, or insurance auditors; insurance loss runs; and electronic mail records.

State and Federal Bureau of Investigation (FBI) Unified Crime Reports

The FBI compiles crime data from across the nation and reports it by city, region, category, age, and other categories. A copy of "Crime in the Untied States" can be found at the local library or at the FBI and other Internet sites.

Bureau of Alcohol, Tobacco, Firearms and Explosives (ATF) and Federal Bureau of Investigation Arson and Bomb Reports

Publications from these agencies contain information on the prevalence of arson and bombing across the United States. The National Fire Protection Association (Quincy, MA) also maintains data on suspicious fires.

Victimization Studies

Victimization studies rely not on arrest or conviction information but on surveys of the general population's experience with crime. See "National Crime Victimization Survey," Bureau of Justice Statistics, Washington, D.C.

Valid Internal or External Surveys

Survey the employee population about its experience with crime in and around the business. Ensure that the results are statistically valid.

Centers for Disease Control and Prevention (CDC)

The CDC maintains information on a range of topics that include violence in the workplace, the use of firearms, and other violent crimes.

Local College Campuses

Although most college campuses are required by law to maintain crime data, crime committed on college campuses may not be reflected in local reporting sources.

Canvassing

Anecdotal information from interviews with patrons, employees, and community members are useful as a double check and can lead the investigator to sources others may miss. Speak with:

- Neighbors
- Competitors
- Fire and ambulance crews
- Union representatives
- Security officers
- Postal carriers
- Regular delivery drivers and suppliers (Federal Express, UPS)

Public Library

Many of the sources listed above can be found at the public library. You can also research past news articles for information at the library or connect through its web site to clipping services.

Surrounding Businesses or Corporations

Examine their incident reports and records and interview longtime employees who would have knowledge of crime, such as the human resources director, security manager, and insurance/risk manager (for insurance loss reports).

Bureau of Justice Statistics

The National Incident-Based Reporting System (NIBRS), Criminal Victimization in the U.S., Violence and Theft in the Workplace, and Sourcebook of Criminal Justice Statistics are some of the useful databases and publications available from the Bureau of Justice Statistics in Washington, D.C.

INCIDENT CLASSIFICATIONS

The following list can be used to help track the occurrence of crimes and incidents. Additional subheadings or subclassifications for crimes committed by employees (internal) or for crimes and incidents committed by customers or guests (external) can be included, depending on the needs of the firm. Robbery can be further divided by including theft by pickpocket (not robbery in some states), purse snatch, or some other division.

Arson
 Actual
 Suspected
Assault
 Simple
 Aggravated (weapon)
Attempted rape
 By employee
 By nonemployee
Burglary
 Attempted
 Forced entry
Computer-related crimes
 Attempted break-in
 Disclosure of passwords
Embezzlement
 Money laundering
 Kickbacks
 Technology transfer
 Extortion
 Forgery
 Counterfeiting
 Misappropriation of funds
Domestic violence "spillover"
Insurance (worker compensation) fraud

Homicide
Kidnapping
    Perpetrator known to victim
    Perpetrator unknown
    Executive or key employee
    Attempted
    Threatened
Theft
    Auto
    Proprietary information
    From auto
    Funds
    Product
    Diversion
    Misappropriation
    Raw materials
    Precious metals
    Personal items
Disturbance
Disorderly conduct
Sabotage
    Suspected
    Product tampering
Vandalism, malicious mischief
    Vehicles
    Tagging, graffiti
Suspicious circumstances
Indecent exposure
Possession or disclosure of objectionable material
Sexual harassment or unwanted advances
Tailgating
Corporate rule violations
Parking violations
Vehicle violations
Vehicle towed
Fire access blocked
Other vehicle code violations
Substance abuse
    Possession
    Sales
    Under influence
Hit and run
Property damage
Robbery
    Strong-arm
    Weapon
    Force/fear

Suicide
    Actual
    Attempted
    Threatened
Access control
    Attempted entry
    Unauthorized entry
    Badge missing/stolen
    Misuse of badge/card
    Loaning access control card
Bombing
    Explosion
    Incendiary
    Threat
    Intelligence/information
Trespassing
Prostitution
Gambling
Gang activity
Alarms
    Security
    Fire
    Environmental
    Process control
Maintenance
    Lighting
    Fencing
    Locks/doors
    Glazing
    Doors/gates/windows open
    Shrubbery/landscaping
Escort requests
Police contacts
Solicitation/special interest
Telecommunications
Fraud
Misuse of company equipment/services
Obscene/harassing phone calls
Demonstrations/picketers
First aid/medical
Safety hazards
Terrorist threat
Stalking

APPENDIX G

Communicating with the Media

Most crisis management public relations disasters result more from a firm's inability to communicate effectively with the media than from the incident that propelled them to public scrutiny. Always be prepared with the facts of the situation when dealing with the media. Their job is to get the facts of a situation quickly and meet their tight deadlines. Although few reporters are hostile, their approach to the type of questions asked often depends on the spokesperson's honesty and sensitivity to their needs. It doesn't matter to the reporter whether the story makes you look good or bad, but a dull story makes the reporter look bad. Reporters will use a variety of techniques to get you to produce interesting facts and opinions. If you volunteer positive points, the reporter is inclined to use them. The alternative is to make the reporter dig out the story. This does not mean that a reporter has total control over this situation. An effective spokesperson may not control the questions, but can control the message through preparedness and by understanding the basic media interview do's and don'ts.

- Do grant interviews as quickly as possible. If you don't, reporters will get their story from people who may offer mere speculation, from a disgruntled employee, or from a worker under the emotional strain of just seeing a co-worker killed or injured. Inaccurate information or perceptions are difficult to change once broadcast.
- Do make sure the reporters know who the spokesperson is. If approached by a reporter, remember that your company may have a policy that only the spokesperson is to answer media questions.
- Do ask the reporters what the story is about, if it is not obvious. Ask if they have talked to anyone else and what the "other side" said in their interview. Follow-up by asking who else they'll interview for the story. Find out how long the interview will last.
- Do establish a warm, friendly, business-like attitude. Learn and use positive body language. Be courteous even if the reporter is obnoxious.
- Do prepare a brief three- to five-message introduction statement. Remember that interviews can be opportunities. Lead with your most important point. If injuries are involved, establish that the company takes safety seriously, or that the public or workers are not in danger (if true). Reassure the public that the company does take customer

service seriously and you regret that someone's mother-in-law won't get her liver transplantation because your on-line auction server went down.

- Do present a consistent and timely account of company policies and activities.
- Do answer: Who, what, when, where, why, how? Questions may include: Who are the victims, what happened, when did it happen, where were victims taken, why did it happen, how many victims were injured/killed?
- Do prepare to respond openly to criticism. Turn negative questions around by giving a positive response. Practice this ability beforehand.
- Do anticipate questions and develop answers. Know your vulnerabilities.
- Do be careful with questions that ask for speculation. The cause of the incident, specific damage estimates, who or what is at fault, or a chronological account of the incident may not be reliably determined at the time.
- Do give yourself time to think by pausing for 1 or 2 seconds before responding. Speak slowly after silent moments.
- Respond to a series of "rapid-fire" questions by picking one that best relates to the point you wish to make. If you lose track of the questions, or become confused, ask the reporter to repeat them one at a time.
- Do keep answers short—use silence effectively.
- Do ask questions that test understanding/acceptance.
- Do be sensitive to time constraints/deadlines.
- Do represent the company in the most favorable light, but speak candidly and accurately. Withholding information that is potentially damaging to the company does not always mean you are not telling the truth. Make sure your information comes from a reliable source. If you need to admit problems, emphasize the positive steps being taken to correct them.
- Do repeat or rephrase questions for an audience before answering (see below).
- Do state your conclusion first when answering a question.
- Do use your own words when answering questions. You don't need to answer at the reporter's speed or tempo, or to sound like a reporter. If you need to think about an answer, do so, but do not delay. Use anecdotes whenever possible and personalize your answers. Avoid referring to "the company." Use "we" or your company's name.
- Do use communications-bridging tactics to change the focus of a question to an issue you wish to cover and away from the topic a reporter wishes to cover.
- Do lead the media conversation.
- Do commit to follow-up with answers.
- Do correct misrepresentations or factual errors immediately, either spoken by you or by the reporter. If after the fact, start with the reporter, not with his or her boss.

- Don't frustrate the reporter's need for basic answers.
- Don't let a reporter interrupt your statement. Wait until the reporter is silent, then complete your thought and continue with any other points that you intend to make.
- Don't use highly technical jargon when answering questions.
- Don't talk to or at reporters—talk with them.
- Don't cite competitors by name if your statement is derogatory, and stay away from liability issues. Don't talk about who is responsible, nor make any accusations. Whatever you say may become part of a legal proceeding, so be as general as possible.
- Don't make financial or project projections. This can have a negative effect on stock prices or other liability issues. Use caution when discussing construction delays, project shutdowns, or other proprietary information.
- Don't expect every media person to be well prepared.
- Don't be afraid to say you can't answer a question. Responses such as "I don't have that information right now, let me get back to you" and "That's the first I've heard of that, let me check it out before responding" are acceptable.
- Don't respond to questions and facts based on unknown sources offered by the reporter. Request to examine the source and facts before you are able to comment.
- Don't say "no comment." The public equates this term with dishonesty and deception. As mentioned above, if the media does not get a comment from you or from the company spokesperson, they will seek out someone who will provide a comment. Most often, this person, such as an employee of the firm, will have nothing to offer but speculation. Rephrasing your response in a manner that avoids "no comment" is acceptable, but phrasing it in a positive manner gets better results. Instead of answering, "We can't make a statement until we have read the legal documents," try "We will be happy to make a statement after we read the legal documents." Other alternatives to "no comment" include: "I'm not the best source of information on this subject"; "We simply don't have sufficient facts to make a meaningful statement"; or "I'm sorry, it's just not appropriate for me to release that information at this time."
- Don't use trigger words from a question in your answer. Don't repeat an offensive question or negative comment even if to deny it. Hostile questions are often cut from the broadcast, leaving you with the offensive quote.
- Don't attempt to influence editorial reporting by promising or referring to advertising.
- Don't give information that is "off the record." Expect reporters to use anything you say. If a reporter wants to talk off the record, say exactly what you'd say on the record.
- Don't wear sunglasses when being interviewed. You will be perceived as hiding something (besides your eyes).

- Don't argue with a reporter or question their motivations. Your angry or hostile reply to an argumentative question may be aired and the question suppressed.
- Don't answer questions defensively or appear to be attacking.
- Don't use statistics if avoidable. If necessary, distribute them to the reporter by printed or graphical means.

APPENDIX H

Security Systems Specifications

[COMPANY LETTERHEAD]

Date _____

Bidder's name and address

_____

_____

Dear Mr._____,

The XYZ Company invites you to participate in the bidding process to provide an integrated intrusion detection/fire detection, access control, and closed-circuit television system at the facilities located at 2727 Sepulveda Street, Torrance, California.

    Attached to this request for proposal (RFP) is the specification that provides the requirements for the system integration. A bidders' conference and a job walk will be held at [time] on [date] at 2727 Sepulveda Street, Torrance, California. Responses to the RFP are due by [time] on [date]. Contract award will be approximately 30 days following receipt of the RFP. Should you decline to participate in the bidding process, please advise me as soon as possible. If you have questions regarding the specification before the bidders' conference and job walk, please call John Doe, Security Manager, XYZ Company, at (310) 555-4005.

Sincerely,

Richard Murphy

(Title)

Enclosure

INTRODUCTION

A *specification* is a detailed, exact statement of particulars, especially a statement prescribing materials, dimensions, and quality of work for something to be built, installed, or manufactured.[1]

A specification for a security system is a part of a request for proposal (RFP) and should provide the bidders as many details as possible. The need for a selection/evaluation team to prepare the RFP is mandatory and should consist of security, finance, procurement, facilities, operational personnel, and other functions deemed appropriate. The selection/evaluation team is normally chaired by security or procurement. The selection/evaluation team jointly prepares the specification, jointly reviews the responses to the RFP, and jointly makes the selection of the successful bidder.

Before issuing the RFP, a bidder's questionnaire should be sent to a select group of suppliers who perform the type of work being requested. The questionnaire should ask questions that are designed to qualify bidders who can perform the desired work while eliminating those bidders who cannot perform the desired work. Information sought may include the size of the company, the length of time in the business, organization chart, financial statement, and references (past and present) where similar work was performed. To aid in the evaluation process, a form should be developed that assigns weighted values to the questions asked in the questionnaire. Another form should be prepared that also assigns weighted values to various portions of the proposal when evaluating the responses to the RFPs. An evaluation criterion includes pricing as well as the overall responsiveness to all elements of the RFP. These forms will aid the selection/evaluation team in the selection of the successful bidder.

After transmission of the RFP, but before the receipt of the responses, a bidders' conference and job walk should be conducted. The bidders' conference provides the bidders with the equal opportunity to ask questions about the project. The job walk provides the bidders with a familiarization of the facility including the location of the system and devices. A record of the bidders' conference should be made to include attendees, questions and answers, and additional information as deemed appropriate.

A transmittal letter is required to accompany the RFP. The transmittal letter identifies the project schedule, including (a) the date, time, and location of a bidders' conference and job walk, (b) the date and time that the responses to the RFP are due, and (c) the expected date of contract award.

The format of the specification will be tailored to the facility and should include at least the following subjects:

Introduction
This section defines the overall system to be procured, including access control, closed-circuit television (CCTV), intrusion detection, or an integration of a number of systems.

---

[1]The American Heritage Dictionary of the English Language, Third Edition.

Scope of Work

The scope of work identifies the physical location of where the work will be performed and the work to be completed by the contractor such as construction, electrical, conduit, systems hardware, software, training, and the supplies to be provided.

System Requirements

System requirements identify how the buyer expects the system to perform as well as any specific requirements that are unique to the project.

User Requirements

User requirements are those operational provisions that the buyer desires incorporated into the system, such as software particulars, specific format of input and output data, hardware system capabilities, and hardware type.

EXAMPLE: REQUIREMENTS SPECIFICATION FOR AN INTEGRATED ELECTRONIC SECURITY SYSTEM

Introduction

This specification contains the requirements for an integrated intrusion detection, access control, fire detection, and CCTV monitoring system for the facility identified in the RFP letter. All quotations must ensure that any inability to comply with these requirements is clearly stated in the quotation submission.

Scope of Work

The quotation is to include (a) the integration of access control, intrusion detection, fire detection, and CCTV; (b) all necessary hardware; (c) photo-identification badges; (d) installation of all hardware; (e) ergonomic console; (f) training of systems operators pertaining to hardware and software; and (g) commissioning of the system as specified including all wiring and equipment. Fire detection will be in accordance with all applicable regulations. The physical location and the number of intrusion detection devices, access control devices, and cameras are contained in the attached drawings. The location and size of the proprietary central monitoring station is also contained in the attached drawings. The quotation should include the contractor's recommendations pertaining to the above requirements.

System Requirements

Intrusion Detection/Fire Detection

The present intrusion detection and fire detection systems and devices will be integrated into a console located at the security control center. The access control, intrusion detection, and fire detection systems will be fully integrated.

Access Control/CCTV

A. The access control system will be a Microsoft Windows–based personal computer (PC) system.

B. The supplier will provide a console in sufficient size to house all state-of-the-art equipment (access control, intrusion detection, fire detection, and CCTV monitors), as well as provide for future expansion. The attached drawings indicate the size and location of the control center.

C. Current light requirements will be taken into consideration when specifying the type and location of cameras. Where required, the supplier will provide the light level requirements.

D. A digital imaging system will be used to create a photo-identification badge. The system should be compatible with generating badges for employees, nonemployess, and visitors in the log-in and log-out processes.

E. The photo-identification badge will incorporate the following characteristics:
1. Colored-coded to designate area access visually, e.g., color background and/or colored bar
2. Company logo on the front
3. Color photograph on the front (sized to be visible from a short distance)
4. Employee's name on the front (sized to be visible from a short distance)
5. Signature block on the rear
6. Name and address of the company on the rear
7. Statement on the rear indicating that if the badge is found, it should be returned to the address indicated
8. Clips utilized to affix the badge to the outermost garment, chest high, on the person. Necklaces, or other such devices, will be provided for use by those who do not wish to use the clip.
9. The badge will include sufficient fields to incorporate personal details, e.g., name, social security number/employee number, department number, card number, access authorizations, and expiration date.

F. The badge will incorporate the following two technologies:
1. Magnetic stripe
2. Proximity

G. The badge technology will be compatible with the existing time and attendance system.

H. Initially, about 1,000 photo-identification/access control badges will be required.

I. For comparison purposes, cameras, monitors, recorders, multiplexers, switchers, quads, or other peripheral equipment will be Brand X or Brand Y. Cameras and monitors will be quoted in color except when physical conditions may require the use of black and white.

J. The supplier will specify all building requirements, e.g., conduit, door hardware, and electrical, as needed, that are not included in this specification.

K. The system will provide for modular expansion of the hardware as well as cardholder capacity for future needs.

L. Access monitoring will be able to detect and report the following conditions: (a) valid request, (b) lost card, (c) wrong time, (d) wrong door, (e) invalid card, and (f) unknown card.

M. The system should include battery backup support.

N. The system database will encrypt the operator passwords to prevent unauthorized viewing.

O. The system will provide for anti-passback.

P. The system will be compatible with other industry-standard office equipment and software programs, in particular the latest version of Microsoft Office.

Q. Door monitoring will include the ability to report door forced open and door held conditions.

User Requirements

Access Control

A. The requirement is for a Microsoft Windows–based compatible user presentation.

B. The system will include comprehensive on-line help screens that relate to the currently active window.

C. A "print screen" command will be required for all screens.

D. System operators will be associated with a log-on password and user ID.

E. The system will have the capability of restricting cardholder by dates, times of day, and reader locations.

F. The system will provide an audible alarm at the console for all unauthorized ingress/egress from any door.

G. The system will provide a detailed audit of the arrival and departure times at any of the card readers. The report will include the ability to sort by any of the personal detail fields.

H. History reporting will be incorporated to provide the ability to review all system alarms, access control activity, and operator actions. Report capability will be through operator's display, printer, or magnetic media. Sort capability will include any of the personal detail fields. In addition, the data in the system will be archived from the system to a digital media to ensure it is preserved.

I. To provide for short-term usage, each card record will have a start and end date validity period. Upon expiration of the valid period, cards will become automatically inactive without operator action.

J. The system will provide a means to back up the system's database.

Closed-Circuit Television

A. Cameras will be pan/tilt/zoom as well as fixed; however, they will be able to provide clear and recognizable images.

B. Cameras and monitors will be color except when physical conditions require the use of black and white.

C. External cameras will be able to view the entire perimeter of the buildings.

D. Cameras are to be installed in weatherproof housings where the elements dictate.

E. The system will provide an audible alarm at the console for all unauthorized ingress/egress from any door.

F. Nine-inch monitors at the control center will be used for each access control point indicated on the attached drawings.

G. A 19-inch monitor will be used for pull-down images from any of the 9-inch monitors.

H. All images will be recorded on a 24-hour digital media and retained for a minimum of 30 days.

I. The number of cameras and monitors is shown in the attached drawings. The location and size of the monitoring control center are also shown in the attached drawings.

CONCLUSION

The specification example contained in this chapter pertains to an upgrade of an existing system and is provided only as a guide. The example is not wholly designed to be applied in every situation. Each facility and each system is unique and should be addressed accordingly. Security systems and devices, as well as related software, are complex, and the state-of-the-art changes rapidly. If the knowledge of state-of-the-art systems and devices and related software is not available within the buyer's organization, assistance should be sought from an independent and objective outside source. Reliance solely on the input from suppliers is not recommended. Procurement specifications also apply to security services such as contract security personnel, alarm monitoring, consulting, and investigations, to name a few. Although the specification for hardware procurements is different from that of service procurements, the process is similar.

Sample Introduction Memorandum: Disaster Recovery Planning

MEMORANDUM FROM IVA BUCKS, CHIEF FINANCIAL OFFICER

Representatives from Corporate Security will meet with key personnel from XYZ Corporation to facilitate the development of a disaster recovery plan by asking a series of questions designed to give them a better idea of your department's day-to-day operations, interdependencies, and the impact a loss of your function may have on the company over time. Senior management considers this a priority project and expects the full cooperation of all participants. Each team leader or department manager is responsible for the completion of his or her portion of the plan, with help from Security. Security employees will provide participants with a questionnaire and instructions for you or your team to list the answers to their questions.

The development of a disaster recovery plan is simple. You will be asked to accomplish the following:

1. Identify your department's critical functions or processes.
2. Determine the most cost-effective strategy to recover these functions or processes.
3. List detailed instructions that implement the strategies and that reduce the need for decision making during implementation.
4. List critical resources needed to implement the strategies.

Critical functions can be defined as a process, service, equipment, or duty that would have one of the following impacts on the company if the function is lost or if access to it is denied:

1. Affect the financial position of the company
2. Have a regulatory impact
3. Reduce or destroy public/customer image/confidence or sales

Functions that can be postponed for a month or longer without suffering the above impacts are generally not considered in the plan.

Recovery strategies can include the following:

1. Transfer of operations to, or increase capacity at, another company site.
2. Contract the work to a third party or competitor.
3. Prearranged alternate work space.
4. Make agreements with vendors to supply preconfigured replacements for equipment within an expedient timeframe.
5. Work at home.

If you have any questions, contact [name] at extension [number].

Sincerely,
Iva Bucks, CFO

Index

Ransom
   crisis management planning for, 231–243
   in kidnap and ransom contingency plan, 317
   payment of, considerations related to, in
         crisis management plan, 238–240
RCRA. *See* Resource Conservation and
         Recovery Act (RCRA)
RDD. *See* Radiological dispersal device
         (RDD)
Reallocation of resources, as recovery strategy
         in business continuity planning
         management, 193
Reanalysis, on resource questionnaire, 218
Receipt(s), security of, questions on hospital
         survey, 276–277
Reciprocal agreements, as recovery strategy
         in business continuity planning
         management, 192
Record(s)
   management of, in mitigation, 115–116
   vital, examples of, 115–116
Recovery
   described, 182
   in response planning
      bomb incident management–related, 141
      chemical/biological attacks–related,
            149–150
      civil disturbance–related, 177
      earthquake-related, 145
      evacuation planning–related, 154
      fire-related, 157–158
      floods and heavy rain–related,
            161–162
      hazardous materials incidents–related,
            164
      hurricane-related, 167
      injury/illness–related, 169
      lightning-related, 171
      tornado-related, 174
      workplace violence–related, 177
Recovery point objectives (RPOs), 188
   BIA's role in, 200
   defined, 200
   described, 200
Recovery priorities, of BCP, 226–227
Recovery strategies, development of, in
         business continuity planning
         management, 187–193
   cold site, 190
   data systems, 191
   equipment rental, 192

   hot sites, 189–190
   manual methods, 192
   purchase of materials from competitors, 191
   quick-ship agreements, 193
   reallocation of resources, 193
   reciprocal agreements, 192
   relocation, 190
   rescheduling production, 193
   service-level agreements, 193
   telecommunications, 190
   third-party manufacturing, 191
   virtual manufacturing, 192
   warm site, 190
   work at home, 190
   workforce management, 192
Recovery teams, in business continuity
         planning management
   setting up, 193–195, 194f
   training, 195
Recovery time objectives (RTOs), 186–187,
         184
   BIA's role in, 199–200
   defined, 200
   described, 200
Redundancies/divergence, as mitigation
         strategy, 111
Regression testing, security systems–related,
         247
Regulatory controls, as mitigation strategy,
         110
Reliability, in cost/benefit analysis, 34–35
Relocation, as recovery strategy in business
         continuity planning management,
         190
Report(s)
   ATF/FBI arson and bomb, as data source in
         crime prediction, 87
   crime, unified, state and FBI, as data source
         in crime prediction, 87
   criteria of, 70–73. *See also* Survey report,
         criteria of
Request for proposal (RFP), 258
Rescheduling production, as recovery strategy
         in business continuity planning
         management, 193
Resident doctors' quarters, security of,
         questions on hospital survey, 277
Resource(s), reallocation of, as recovery
         strategy in business continuity
         planning management, 193